Surveillance,
Transparency,
and Democracy

Public Administration: Criticism & Creativity

Surveillance, Transparency, and Democracy

PUBLIC ADMINISTRATION
IN THE INFORMATION AGE

Akhlaque Haque

THE UNIVERSITY OF ALABAMA PRESS
Tuscaloosa

The University of Alabama Press
Tuscaloosa, Alabama 35487-0380
uapress.ua.edu

Hardcover edition published 2015.
Paperback edition published 2020.
eBook edition published 2015.

Inquiries about reproducing material from this work should be
addressed to the University of Alabama Press.

Typeface: Minion

Cover image: © Jokerproproduction | Dreamstime.com
Cover design: Kyle Anthony Clark

Paperback ISBN: 978-0-8173-5988-1

A previous edition of this book has been catalogued by
the Library of Congress.

ISBN: 978-0-8173-1877-2 (cloth)
E-ISBN: 978-0-8173-8876-8

Contents

Figures

Preface

Information technology is an integral part of public administration. Millions of people depend on the government on a daily basis for services that range from garbage collection to national security. The most basic of human needs, including food, shelter, and health care, are administered by government entities. Such service provisions are intrinsically tied to how information is managed, used, and disseminated through information systems and various information and communication technologies (ICT). These tools are not only changing how governments interact with citizens, they are also changing the priorities of governmental tasks as administrators battle with fiscal austerity and evolving governmental reforms. ICT can make the difference between whether a social worker should respond to continuous tweets and update blogs or attend a court battle for child custody. Likewise, a senior municipal planner may have to make the choice between whether to prepare a report from numerous databases for a competitive grant or go to a city hall meeting and respond to citizens about dwindling streets and zoning violations. Ultimately, ICT will fundamentally change the role of public administration in a democracy.

Administration organizes information and knowledge around a defined purpose. Practice, on the other hand, applies information in real-life situations. There is a wide disconnect between administration and practice—between what we *know* (organized thought or information) and what we *apply* in real life (practice) from what we know. For example, we *know* we don't have the money, yet we overspend despite the odds of becoming completely broke. We have poor health, yet we tend to eat food that further harms our health; we hear but we do not listen; we are informed, yet we remain blind when it comes to applying that information to practice. Commenting on the state of the field and our "growing relative ignorance" about public administration, Todd La Porte (1994) summarizes: "Our grasp of the dynamics and behavior of public organizations is slipping further and further away: *we know less of what we need to know, even as we know more than we did* —and even as prescriptions for change and improvement proliferate" (La Porte, 1994, p. 7; emphasis in the original).

We generate megadata nonstop at incomprehensible speeds, and as a result, administrators fall under the avalanche of data in the frantic futility of storing, maintaining, sorting, arranging, interpreting, and reporting data. Time and manpower are expended, but how much of that fruit reaches the citizens' market? Technological development and administrative practice are out of step with each other, with the result that even the most valuable new information found in practice may not find its way into policy directives. When we are tied up in administration, we choose data for answers, rather than reflecting and learning from our practices. This is the practice-administration dichotomy in the Information Age.

Modern technology presents us with an obvious challenge and a dilemma: Should we enact information technology merely to administer government programs, or should we use ICT to learn from our practices? Technology enactment exists within a social order. It is not independent of how society perceives it. Communities have long been segregated by means of physical walls, roads, and artificial boundaries. Today, communities are segregated by information systems, databases, Internet clouds, and various standardized archives, either coded as individuals or labeled in groups in different forms and formats. Humans are trapped into information surveillance systems as data bits, as if waiting and ready to be used by an omniscient researcher or a private entrepreneur who will tell them how to lead a better life or, at least, tell them the next day's weather.

Public administrators use ICT primarily for administration, processing, and control of information. With increasing technological supremacy, government and businesses are engaged in a shopping spree of collecting data. The flood of data coming at lightning speed sets a momentum and demands attention. Data begs to be monitored, managed, and manipulated. These time-consuming tasks burden the administrative capacity of the government, supplanting the primacy of the people the whole apparatus was meant to serve. This constant engagement with technology and data can become an obvious challenge for government when its administrators know more about organizing data than about how to convert its value into practice. The National Security Agency's citizen surveillance goals should remind us that the administrative goal of using technology for gathering private data is not always compatible with the practices and values of a democratic nation. Edward

Snowden's revelations about data gathering efforts by the NSA may have surprised the world, but the ease with which data can be captured makes privacy a nagging nuisance. Indeed, the science of data collection is much more attractive to system administrators than the art of managing and protecting the data. "Big data" generates economic value and the opportunity (and the upper hand) to customize services based on citizens' profiles. It is not surprising that telecommunication giants are being paid millions of dollars by the NSA through the "black budget" to have access to citizens' information (Timberg & Gellman, 2013).

There is significant interest among public and private entities to capture competitive advantage over data collection. Having background knowledge of citizens/clients gives providers greater control in managing resources in terms of understanding citizen needs and cutting waste (McGuire et al., 2012). The upshot of the data gathering efforts is that it has created new administrative machinery involved in the business of *citizen commodification* by depersonalizing (or value neutralizing) data. Furthermore, the increase in data sharing arrangements among governmental bodies and the private sector is blurring the mission of public agencies as it threatens trust in government.

Although ICT on a grand scale was first in the hands of public administrators who used their capabilities to process and control data unilaterally and top down, the Internet presents a greater opportunity to capture the *values* of citizens through social and political engagements. According to Pew Research (Fox & Rainie, 2014), 87% of all American adults are Internet users, with 97% users among the Millennials (those born in the early 1980s to early 2000s). The constant online presence of citizens provides unprecedented opportunity for public administrators to share the practice of government. As much as public administrators want to focus on data for improving public service delivery and safety, they inadvertently become antagonists who appear to have the higher power to pry into citizens' business. The traditional one-way usage of technology *dictates* that citizens be engaged with government rather than *inviting* them to participate. This consequently disengages the citizens, and the ability to learn from practice is lost or limited.

Whereas data-based learning connects an administrator with data and tools, practice-based learning connects administrators to citizens. Often unseen and unheard is the knowledge embedded within administrative practices. Administrative practices are the everyday, routine business of government. Practices embody the value of government as citizens interact with administrators regarding services and the implementation of law and order in society. Within those practices are powerful stories about the workings of government and interaction with citizens. Whereas data is primarily value neutral, practice is the explication of values within a context. When the knowledge of citizens is limited to data, the valuable knowledge of practice is lost, and citizens are reduced to some quantifiable data input. When millions of people in the United States could not get access to the new healthcare.gov website to sign up for health insurance, it became clear that the practice of government is a much more complicated business than ordering a couch from Amazon.com. It became clear that citizens needed access to the online world before they could get access to health care, yet vast numbers of the uninsured had neither the skills nor the equipment to connect with the new required insurance. However, to many practitioners it was not surprising that healthcare.gov failed to provide promised services on time during the October 2013 implementation of the website. Only about 6% of large-scale IT projects are successful (Gross, 2013). Millions of dollars of IT projects are abandoned every year because programmers and planners who design the projects make faulty assumptions. IT experts, accustomed to a top-down and one-way administration, abruptly discovered that tested knowledge from practice, from the consumer up, would have been valuable. Even during the sign-up of uninsured clients to the newly created health exchange or insurance marketplace, the "navigators" and health care counselors have been far more effective in helping the uninsured than the healthcare.gov website. Today, practice-based learning is often lost or even undermined by the onslaught of the waves of data that are being administered by all governmental agencies. Whereas data-based learning is value neutral, practice-based learning incorporates the values gathered from reality on the ground. Unbridled use of science and reason is likely to trample practice-based learning. More importantly, overreliance on technology's ability to solve social problems in isolation will result in decisions that disregard political values, the values of the very people directly affected by such decisions.

What makes ICT particularly useful is its ability to process information quickly. However, what would make ICT truly revolutionary in a democratic society is its capability to process information without corrupting the original intent of the transmitter, so the receiver has control over the information to interpret it according to his or her knowledge base. Academicians and practitioners in the field have yet to ask the tough questions about processing information and how the processing impacts democratic institutions and the constituents they serve. How information is produced, reproduced, and represented through information systems should be of concern to public administrators because it transmits values that affect everyone.

The decision to use information gathering and processing technologies depends on the broader goals of the reflective administrator who continuously learns from practice. The values of citizens, priorities of the community, and the return on investment measured in the ability to meet the needs of the citizens all inform the decision an administrator must make. Experience and knowledge learned from practice are essential for one who must weigh all the elements. For all its benefits and advantages, information technology does not govern; it is a tool that demands wise judgment, planning, application, and ethics in every phase of its design and application. Experience means more than time and proficiency in the office. It is the proven ability of the administrator to assimilate capabilities enhanced through technology to learn the values of citizens and to respond to them more effectively. Experience must precede new technology initiatives to ensure that information gathered from practice drives technology usage—not the contrary. At this point in development of ICT sciences, many are infatuated with technology. Enthusiasts laud it as though it had a life of its own, a magical ability to solve all our social and political problems. Isolated from practice and ignorant of the needs and wants of other citizens, ICT today has created a false sense of confidence about public organizations making good decisions based on the sophistication and processing capability of software and hardware. Chasing the myth that more sophisticated information technology will reduce uncertainty, organizations are led to invest further in technology, including technologies designed to monitor their citizens' activities. Therefore, it should be no surprise that despite significant e-government initiatives, most e-government projects are unable to have meaningful impact on governmental reforms because information from practice is undermined in the face of technological supremacy (Heeks & Stanforth, 2007).

The Role of Technology

Humans create the meaning of any data from practice—how it is used, for what purpose, and who is likely to use it. Information systems, on the other hand, do not create any new meaning. Instead, information systems process information to create, on demand, a summary of discrete data that is ready for interpretation. The summary is a snapshot prepared for a targeted audience whose interpretation of the summary is already implied, given the selection of inputs that generates the summary. For example, I can take a few variables from one of the 385,000 datasets from the data.gov portal (which houses the largest online public data in the United States) to create a graphical summary in an Excel worksheet that shows housing conditions in an area. Given the limitations of technology, the reproduction of the data in the form of graphical representation will be greatly influenced by how well the Excel data sheets were prepared and how convincingly the analyses were communicated. When technology is involved in information processing, it changes the ontological nature of the information. In other words, information systems ascribe meaning to the codes entered into the system based on a particular worldview. Therefore, information systems reflect the human biases of their designers in the outcomes generated by the said systems. ICTs in general are preconceived for a desired outcome as deemed useful for the audience in question. Large-scale e-government initiatives for collecting taxes or providing services and small-scale data management software have their own intended audience and purpose that are not independent of the social values that manage and control them. It is incumbent on the institution to be cognizant of the limitations of ICT, including large-scale data gathering schemes, to make policy decisions. This will require public institutions to develop an intentionally refined sense of social purpose and a much more cautious stance when implementing technology (La Porte, 1971). Most technology implementation today is debated primarily by IT experts or oversight committees whose knowledge of controlling the unknown is limited to what the technology is meant to do—whether it works, or how many times it fails. The people who are expected to benefit from the technology are given only the experts' opinion about its potential. Thus any uncertainty surrounding the scheme remains untested and is completely out of the hands of the public who will be affected by it. With this in mind, we can argue that new technology may be implemented while going unchecked by any democratic involvement, and subsequently harbors the potential to conflict with the higher value of protecting the public interest.

It is high time that we reevaluate the administrative perspective on technological applications in order to reduce the tension between administration and practice and to pursue liberating political and human values. In order to control the consequences of their own actions, public administrators must be educated about the broader role of technology so they can create new avenues of social and political engagement. Administrators, those tasked to take the goals and requirements of citizens and convert them into actionable, real practices, must participate in the development to invite citizen engagement. Technology today is not just a tool, but part of a larger process that has implications beyond office management and far into shaping social and political order. Just as elected officials can blame public administrators for unexpected consequences of their policy mandates, in turn, public administrators can blame the tools for failing to see the train wreck before the disaster strikes the bystanders (the citizens). The citizens become the collateral damage in the actions of government and public affairs.

This book explains the practice-administration dichotomy in the Information Age. The subject matter discussed here resonates with scholars who have viewed technology as an active part of the social world affecting the informal processes of an institution and individual practice. Technology, as an enabler, when enrolled with a desired outcome, goes through a process of translation with human actors. Human actors initially play an active role in controlling the information systems' outcome because of uncertainty of what it will produce. But once the outcome is routinized to produce the types of information that suit the interest of the user, the automaton performs without a thought. Hence, technology, in general, is part of our social order exactly as—and only as—we define it. I explain the critical need to articulate the proper use of technology in society so that public administrators and citizens alike are aware and involved in transforming technology from a monitoring and surveillance tool to an agent that enhances information transparency and facilitates learning for human development. The theoretical construct developed in this book will help institutional leaders and public administrators make better decisions using the human and nonhuman means available in the modern technological era.

This work seeks to fill a void in the literature dealing with the role of information and information technology in government. For example, it adds to Jane Fountain's *Building the Virtual State* (2001), addressing the misconception that information technology has no predictable influence on institutional change. Rather, as Fountain argues, a large part of technologies' influences on institutions depends on institutional norms, individual cognitive patterns, cultural elements, and the worldview of the institutional members. Although her argument clearly sets the stage for a larger discussion of the impact of IT on institutions, there is much room left to discuss how technology can be an effective (and democratic) vehicle for organized institutional reform. Whereas rational organizations seek to graduate from organizations to natural institutions by adapting to external and internal societal goals (Selznick, 1957), technology without proper guidance disturbs the settling process. As technology proliferates, it reinforces the surveillance mechanism already in place and becomes an integral part of a more sophisticated rational system.

I argue, following the rational bureaucratic model, that public technology is currently used to feed the demands of the existing institutional framework that constrains the discretionary authority of public administrators and undermines the democratic values of a pluralistic society. Failure to use information—the primary ingredient of IT—in ways that benefit society will not only make the institution "hostage to its own history" (Selznick, 1992,

p. 232) but also make the public institution less valuable to its own citizens. It is not surprising that even when we are drowning in information, we are unable to make the right choices because we are surrounded by institutional walls. As argued by Archon Fung and his colleagues (2007), "In the United States, that prides itself on openness, secrecy becomes a closely guarded privilege" (p. 172). This work joins many scholars who have welcomed information technologies' capabilities for an empowered citizenry, with the hopes that mere efficiency gains from faster information capability do not supplant the desire for a stronger democracy.

Acknowledgments

This work owes thanks to the endless combination of ideas accumulated from the works of many minds from various fields of study including public administration, sociology, behavioral and institutional economics, psychology, geography, anthropology, neuroscience, computer science and communications. In particular, I owe a great debt to the contemporary philosophical tradition that has significantly influenced my work, particularly through ideas expressed in the works by (in alphabetical order) Andrew Feenberg (b. 1943), Harold Garfinkel (1917–2011), Anthony Giddens (b. 1938), Martin Heidegger (1889–1976), Bruno Latour (b. 1947), and Philip Selznick (1919–2010). Perhaps my interest in society and democracy bears its roots in the works of Edmund Burke (1729–1797) where I started my journey to explore the history of ideas.

I am indebted to my students, colleagues, and friends who continue to inspire my passion for learning. I owe a great deal to my dear wife Saira and our children Sabirah and Amaan who have been patiently waiting to see this work completed. I hope in this endeavor to reflect admiration for my deceased parents who will never know how much they have inspired me to do this work.

Surveillance,
Transparency,
and **Democracy**

Introduction

The World Wide Web (WWW) was born in March 1989, when Tim Berners-Lee, a research scientist with the European Organization for Nuclear Research (CERN), wrote a concept paper about how to minimize information loss within his organization. Berners-Lee's initial thought was to solve the information management problem of how to retain and reuse the information that was being lost due to scientist turnovers at CERN, and due to the hierarchical treelike classification of information management systems. The birth of the WWW came about when Berners-Lee was able to lay the groundwork for an information architecture that would preserve the integrity of an efficient information management system and allow new information to be created and managed by individual entities. The system would also have the capability for the new information to be shared with the social world. Berners-Lee's idea was a perfect marriage between the technical world of administration (of network protocols) and the social world, waiting to learn from information and discover how to improve people's lives and the human condition. Berners-Lee (1989) argued that "the aim would be to allow a place to be found for any information or reference which one felt was important, and a way of finding it afterwards. The result should be sufficiently attractive to use that it [*sic*] the information contained would grow past a critical threshold, so that the usefulness of the scheme would in turn encourage its increased use" (conclusion, para. 1). The concept that any information could be alive and floating for others to retrieve even after its immediate need has been met opened a new paradigm of learning and relearning. The dual nature of the WWW, with its formal rules of operation through network protocols and informal liberating power through human networks, makes it one of the most potentially influential tools for human development in the modern world. The vision of the WWW also attests that efficiency gained through technology and technique needs to be open to diversity of values so it can seamlessly integrate with the society it ought to serve. The concept of the WWW has important implications for understanding the relationship between information that is used for administration and applied into practice.

Scholars have long argued that public administration, as a field, suffers from an identity crisis because there is no coherent understanding of the role of public administrators who serve multiple masters and interests. In general, however, it is established that public administrators play a dual role in government. They must maintain a balance between providing policy guidance to elected officials while administering the implementation of those policy mandates based on broader goals envisioned for the citizenry. As individuals, public administrators (by design) act as the medium through which elected officials serve the citizens. On one hand, elected officials expect public administrators to carry out duties with clear and standardized objectives as envisaged in policy directives. On the other hand, citizens expect that public administrators will execute their duties responsibly with minimal economic and social hardship to the citizenry. The Friedrich-Finer debate (1935–36) reminds us that the task of the public administrator becomes increasingly like a balancing act with respect to being relevant to public purposes, as well as being relevant to the elected representatives. Thus the extent that public administrators are *informed* about the society they serve and how well they master the expertise to do the job are critical for successful execution of public service duties. Therefore, in order to fully develop the required characteristics of a functional public administrator, each administrator must be equally *informed* about the society they serve, and about the intricacies of the job to be performed. The "thinking" aspect of the broader goals, expectations, and values of society and the "doing" aspect of using skills to perform a task are inseparable. Thinking of what ought to be the social goals and doing the work to be an effective functional body of the government are the simultaneous tasks of the responsible public administrator in a democratic society.

The information that aids individuals in thinking about social values is quite different from the information that shapes people's understanding about how to perform a task. Simply stated, information can be categorized into two groups: one being *how to do the job* and the other being *why do it?* Together these two groups shape our understanding about *how to act* in a given situation and *whether to act*. The two are of equal weight, and both types of information must be accessible if one is to be engaged beneficially in the affairs of society and government. Without being equally informed by both types of information, one will clearly limit one's ability to perform responsibly in a democratic society. One important distinction between the two types of information is that how-to-do information is *formalized* so that it can be replicated as "tools" for others to use it. However, the why-do-it information is *informal* and open to individual interpretation.

Duality of Information

Facts can be described in technical terms so that the information delivered is unambiguous to the extent that it is quantifiable. This is the formal information that is standardized by minimizing the noises associated with the information. A simple example would be to describe a cup of coffee. The information can be described just as a regular cup of coffee or formalized to 8 oz. of coffee with a temperature between 165 to 175 degrees Fahrenheit. Informal information, when not formalized, gives the receiver a wider discretion to interpret the meaning based on his/her understanding of the object being described and its relation to other related phenomenon. When it is described informally, a cup of coffee could mean more than just the coffee itself; it can include the hospitality and/or the environment in which it is being offered. The father of the mathematical theory of information, Claude Shannon (1916–2001), explains that formal information is more of a signal or symbol that *carries* the information but not the information as a whole— it is the data without the information. Formal information is the study of information at the syntactic level. Since computers are syntactical devices, formal information is appropriate for communicating with computers, including information and communication technology (ICT).

The fact that ICT is able to capture only one part of information (formal) narrows its capability of transmitting the whole information so that the receiver of the information can make the best use of that information to interpret reality. This limitation was first noted by Joseph Goguen (1992, 1997) who argued that "dry" and more formal information has to be reconciled with "wet" or informal information in order to fully comprehend the essence of the information that is translated from humans to information systems. Goguen argued that the theory of information should be a social theory of information as opposed to a statistical or mathematical theory of information (as in the works of Shannon & Weaver, 1964). He held that information is not value neutral. In order for people to transmit the value of information they must be sure the recipient understands it. This is what is meant by reflexive. A message was sent, but was it received whole? If we rely on information systems to relay the message, it will only be partial information because the part of the message that will be transmitted is the information that can be captured by the systems. The transmitter relies on the system to send a particular message and expects the receiver to receive that message without distortion. However, how well the receiver gets the message will directly depend on how reliable the medium is to capture that message and deliver it to the recipient. Citing the work of Harold Garfinkel's ethnomethodology, Goguen argues the types of information used for designing information systems matter because they are the embodiment of the values that will be generated from the systems. He argued that to develop socially desirable information systems, we must pay close attention to what distinguishes formal or dry information from informal or wet information. Formal information is noncontextual, standardized, and has a universal language for communicating among peer groups. Information systems are built to capture formal information because they are objective facts explained through numbers or value-neutral codes. Formal information provides the basis for developing methods of engagement and solving problems because they are targeted to explain how to do a job, how to organize a budget, or how to kill a bird. Informal information, on the other hand, is contextual, unorganized, and grounded in local values; it emerges from social interaction and is localized by the place and language where the interaction takes place. It carries different meaning to different people given conditions of time and place. It is not meant to be used for explaining how to do things, rather why do them. Informal information in this regard should carry equal, if not greater weight when that information is used to serve others. This is because public service requires an understanding of the values of the people who are being served, not merely an understanding of the values of the people charged with the responsibility to serve.

The implication of dry and wet information is far reaching for contemporary public administrators in the Information Age. At the individual level, thinking and doing occurs simultaneously rather than independently. Historically, however, the two types of information have been kept separate by scholars in public administration as formal information continues to influence and dominate the discourse among many mainstream scholars in the field.

The proponents of scientific management principles have argued that because of the efficiency gains from division of labor, thinkers should be separated from doers. One group will primarily "think" and plan, and the other group, the "doers," should be dedicated to executing the plan. In other words, one group will be responsible for doing things as long as the thinkers tell them how to do them. Formalized and targeted information will be critical to the doers for successfully completing the task, given that the thinkers have already decided through informal information as to why they are doing it. This led to what appeared to be a rational approach—to systematically separate the two information categories through the established protocol. All that remained for the management theorist became simple: develop an effective coordination mechanism between the thinkers and the doers. One problem with this approach is that because the groups are separated by design, information is also selectively chosen for each group. Information considered important for the designers of the tasks would not be necessarily the same information made available to people expected to complete the task. This separation of information is beneficial and highly effective in business management practices where the thinkers strategize what to produce (the ends) and the doers are asked to find the best means to get the job done. In principle, once it has been decided what is to be produced, the task of how to produce, maintain, and monitor becomes, by default, the job responsibility of the administrator. By artificially separating the information entrusted to each group, the rational theorist limited the scope of the information resource and failed to use it effectively for societal advantage. Formal information that is also standardized in order to give a universal message fulfills the requirements for giving orders to workers for completing the task. Informal information is reserved for thinkers who use it to make value judgments and to reconcile the conflicts in decision outcomes. In such a model, the contribution by the lay actors in producing information through practice, particularly through their social interaction with citizens, is fully ignored in the development of "thinking." The approach diminishes the overall value of public service.

Information is fundamental to a viable democracy. Because information is the primary resource for making informed decisions, it is also a right for every citizen. Making artificial subdivisions within information not only misrepresents the actual information but also misleads those who depend on that information. Within the rationalist framework, informal information is a formal requirement (perhaps more of a privilege) reserved for the thinkers that the doers are systematically denied. Such practices can be very effective in for-profit organizations where the value of the market is judged by the thinkers to make strategic market decisions to stay competitive and also to increase return on their investment. For public agencies, however, such practices are not only counterproductive but are also undemocratic. Just as thinking about social issues (that is, "market condition") is a duty of the public administrator (Frederickson, 1980), so is "doing" to manage the public business and to get the job done as citizens expect (Kettl, 2002). As scholars in the field, we have devoted significant attention to the question of how to do the job, but not enough attention to the why-do-it aspect. Formal information is far more appealing because it is objective, quantifiable, and prescriptive in nature. Hence, it can be universalized and ready to be transmitted beyond time and space. Information that arises from the practice of government generates the answers for the "why" question of public service. Because such information is informal, value laden, and dynamic, it is difficult to capture. It is also difficult to transmit informal information via an information system because the "whole" cannot be broken into parts through syntactic codes and therefore remains unsaid, or at best ignored. It is convenient to settle with formal information when information technology is available. The professional's sheer love for finding technological solutions and techniques from formal information is also one of the primary reasons why practice-based learning is less appealing to practitioners.

The ubiquitous presence of information technologies has heightened the need for formal information so it can feed systems with billions of data points for answers. There is more data available under our fingertips today than at any time in human history. Yet under the tsunami of data, we are unable to make the best use of the available information. This is because we are dealing with billions of formal information or data points that are disconnected. In order to make sense of the data points (that partially represent the whole information), we would need the help of the informal information gathered from practice. Informal information can explain the relationship between discrete data points, thus making sense of the dry information. Now, to a growing extent, we have artificially separated the formal and informal information for our convenience, and information technology has only exacerbated the separation of the two types of information. Although we are practically submerged in practice-based (wet) information, we are unable to navigate the waters. This is again due to the fact that in order to make decisions, practice-based information is practically ignored. Being intoxicated by objectivity (quantification) and precision, we expect formal (dry) information to carry meaning that is predesigned and targeted toward a given solution. The receiver is only *taking orders* as technology facilitates this efficient information management enterprise.

Indeed, formal information cannot be separated from informal information to make sense of the whole. In other words, thinking (informal) and doing (formal) should go hand in hand. If ICT is to be the primary medium to transmit and process the information, we will inadvertently give precedence to formal information. To find the balance we must discover how technology can be useful to both formal and informal information.

It is important to note that how much formal information will sway an individual to make decisions varies. The reconciliation of formal and informal information happens at the individual cognitive level depending on one's level of social interaction (Rabassa, 2005). More social interaction leads to the establishment of *relationships* from experiences, fostering higher cognitive ability and the confidence to deal with the unknowns. This is possible because individual *relational experience* opens avenues to connect one's thought to action. People having limited experience (that is, limited social interaction) will have a harder time connecting thought to action, even though they might have a lot of experience dealing with machines and nonhuman artifacts. Generally, people having more interaction and exposure with people will have a larger pool of information from which to draw. Contrast this with the resources at hand for people with experience predominantly with data and machines (that is, computers). Those people might receive a lot of formal training and information but may not be able to deal with real-life contingencies in society unless the formal training can be reconciled with social experience. Therefore, we can expect that people's decision making will be influenced by how much formal and informal information they have gathered in their lifetimes.

Practice-Administration Dichotomy

It is time for the Information Age to come of age, to mature beyond the blind infatuation with an unprecedented ability to produce enormous volumes of data. Merely generating such data has been intoxicating, giving administrators mountains of information that can block the view of purpose. When data rules decision making, public administration faces a fundamental choice. Should we seek and tweak more of the formal information (that is, data) and develop and refine scientific techniques to accomplish that task, or should the focus be on discovering among the citizens what problems are immediate or emerging? Did the tasks we accomplished solve urgent problems and lay the groundwork for future complications? If not, why not? The choice is clear; should we—"steer or serve"? (Fredrickson, 1990; Denhardt & Denhardt, 2000). In other words, whether we invest in the "how" factor— how to get the job done—or we invest in the values of service—"why" we do it, I argue that we do both. However, given the dominance of the rational-standardized approach to decision making in public administration, and that information technology favors developing techniques rather than reflexive action, I argue at this point many steer more than they serve. To earn citizens' trust, we must refocus our attention to serve by establishing relationships with citizens. Mutual understanding develops the emotional bond that is fundamental for establishing trust between citizens and public servants, particularly if all are to act morally (Harmon & McSwite, 2011).

In the context of a public administrator, to be informed is to be able to use the information necessary to become socially mobilized to act on information in order to have an impact on the lives of citizens. To reach out to citizens, modern-day information technology can be an important tool if public administrators are cognizant of the limits of technology and fill the gap of informal information to *act from a distance* where technology has yet to traverse. Therefore, I argue that to be able to use information technology for social advantage, we must take a bird's-eye view of where technology fits in our social milieu. The constitutive properties of technology should help public administrators explore the possibilities of using the democracy-enhancing features of information technology.

I borrow ideas from actor network theory (ANT) to explain how both formal and informal information can be reconciled to create the relational value of public service through modern-day information technology. The relational value is created by our understanding of society as an assemblage of things and people under a sociotechnical framework. Human and nonhuman actors, including technologies, are associated with each other to form a network of association to accomplish social goals. Where humans fail for lack of processing power, technology comes as an aid. Numerous networks are part of our lives as each of the human and nonhuman actors are actively *engaged* (even at the subconscious level) to accomplish various social goals, starting from raising a family, to keeping a job, to becoming part of a community. In this assemblage there are two critical parts that sustain the networks: (1) The technical idea that fulfills the technical requirements of the mobilization and, (2) The practice by which values emerge to become part of our routine. Technical information is formal information with standardized objectives clearly identified to perform a task. Practice is habitual and routine, embedded within information. Technique and practice are independent of each other but continuously communicate within the social association. Whereas technical ideas are noncontextual, based on objective facts and formal rules, practice is contextual, local, informal, and often ambiguous. The technical idea informs us how to connect or accomplish a task. Practice, on the other hand, tells us why certain technical ideas are important, and tells us which techniques perform certain acts that fit well with our routine. Furthermore, whereas technical ideas are based on values that are given (established by individuals involved in devising or negotiating the technique), practice-based values are ever changing as issues emerge and as people and things associate themselves in different configurations.

Practice-Administration Dichotomy: Conceptual Framework

Public administration is a practice-based field. Practice and experience are the learning "tools" that can prepare and guide public administrators to become better managers. Whereas to *administer* is to implement and monitor the regime's values (Rohr, 1989), to learn from *practice* is to enable new values to emerge, to be evaluated and perhaps incorporated into evolving democratic ethos (Cooper, 1990). To administer in a democracy is first to understand emerging values, and then to act on those values (Frederickson, 1997). Although tension between administration and practice may be found in a democracy, the larger philosophical underpinning has not been very well understood. The role of public administrators is beyond administering to becoming an active and informed citizenry. Information plays a dual role in informing public administrators on how to manage public business and also answers the why question of managing the public.

The dichotomous nature of information has direct implications for administration and practice. Administration is defined as the performance and execution of responsibilities. It is prescriptive in nature as to how to perform or how to execute. Under bounded rationality, it requires an organized way of solving a given problem. Formal information is most useful in giving instructions on how to perform a task. It is formalized so as to give a structure and targeted toward a given end. Practice can be defined as habitual performance. It is learning about why certain things are done the way they are done. The information required to understand practice is informal, descriptive, nonstructured (without a form), and is open to interpretation. The type of actions we exemplify can be traced back to the type of information we use to make decisions. When we are being evaluated on the performance of a task and given detailed formal instructions on how to do the job perfectly, we may operate as a machine without questioning (or knowing) why we are doing it (other than getting a better performance evaluation). When we see others doing the job, we can relate to the broader picture of why the task is being performed and the value of the work being performed (how it benefits citizens, and such). The duality of the relationship between informal and formal information is reflected in the practice-administration dichotomy. In other words, the dichotomy between administration and practice can be explained with the type of information one uses to be informed and then acts on. This is clearly the case when someone is informed to be an *efficient*, get-the-job-done kind of public servant, and someone who is informed to *serve* the citizen interests.

In figure 1 the practice-administration dichotomy is explained using a visual framework that illustrates the dual nature of information and its important implications for administrative actions. The public service information framework factually explains that the value of public service is wholly dependent upon the types of information that is valued, and how it weighs into decision making. The type of information required to make a decision about "how to do" a public service task is fundamentally different from the type of information required to answer, "why do the service?" Together, these two types of information define the value of public service. When we include both types of information into our decision, we not only know how to do it, but also why we are doing it. This allows us to be reflective in our actions so that in making decisions, moving forward, we use constrained discretion instead of unconstrained or indiscriminate discretion. The left side of the diagram in figure 1 relates to the formal side of the information that informs public administrators how to implement policy mandates. Information technology, and technology in general, plays a dominant role in mediating administrative practices toward automation. The epistemological assumption is that the formal rules of administration can be reproduced by machine. Such tendencies are more prevalent in agencies where more formal and noncontextual information dominates the culture (the Department of Defense, for example). Automation is essentially creating a "black box" or closed loop system (Kaghan & Bowker, 2001) that is set up to connect humans to nonhumans (computers/machines) through standardized protocols or intermediaries (Latour, 2005). The "black box" can be defined as a technique with identifiable input and output to get a job done. It is a system that has been rationalized to get an identifiable objective ex ante. Large-scale information management systems such as the enterprise architectures that domi-

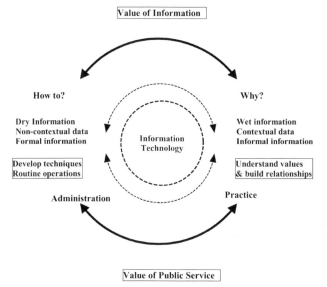

Figure 1. Public service information framework

nate modern-day information systems are essentially created to institution-alize the information automation.

The right side of the diagram focuses on the informal side of the information that informs public administrators about the context where policies will be implemented. The informal information embedded within the culture has to be understood from social experience. The information is dynamic, and it demands constant thinking about administrative goals given the particular policy mandates that require immediate action. Information technology can play a critical role in becoming a mediator, as opposed to an intermediary, when and if administrators become active participants through social media and *act from a distance* (Latour, 1987). Intermediaries are "information products" or transmitters; they do not add value to the information. On the other hand, mediators are the enablers that constantly negotiate with humans and nonhuman actors to understand the relationship between various actors. The interaction with actors (that is, the practice) over time helps solutions to emerge rather than be assigned deliberately from a predetermined set of solutions. The quintessential value of public service emerges when we are able to reconcile thinking and doing by focusing on developing ontological solutions to both sides of information at the same time.

Public Administration in the Information Age

The tension between administration and practice has not received due attention. When public administrators devote their energy solely to administration and neglect communication in practice, they lose connection with their constituents. Then government appears nonresponsive to citizens' aspirations. Ultimately the price is paid; citizens start losing trust in their government and make irresponsible choices that further alienate citizens and governments. Therefore, the role of public administrators is beyond administering to becoming an active and informed citizenry. Here, information plays a key role in informing public administrators how to manage public business and provides insight to the "why manage them" question.

Two scenarios are presented below to explain the tension between practice and administration in the Information Age:

> David Farmer is the new senior comptroller in the state accounting office. His primary responsibility is to monitor the balance sheets of the state's numerous departments and ensure their compliance with regulations. After three months' training in Comp-Solutions software, he is ready to work with the program that large government and business en-

terprises use to track compliance with regulations and cost accounting standards. The software helps to determine if funding is available and if departmental spending complies with department and state budgets.

The major part of David's work is to administer the data submitted by each department through a secured Internet portal on the state accounting office's system server, where the software is installed. With the help of two staff members under his supervision, David must protect the integrity of the data and manage details of each department's budgetary compliance. He reports on this compliance to his supervisor every day. The software is the information processing unit David and his staff use as an administrative tool to get the job done.

David recognizes the importance of his purely administrative work if the state is to maintain fiscal stability. He comprehends that any breakdown in the machinery can have serious consequences for his job performance. Smooth functioning of the information system is essential to his productivity. Moreover, the capability of Comp-Solutions software itself is central and critical to the functional integrity of the entire state's budget. David's confidence is based on trust in the software program and the processing unit.

Michelle Warden is a public administrator at the county level. She is director of economic and community development. As a senior manager for the last 15 years, her primary responsibility is to oversee funding through a multimillion-dollar community development block grant. The funding provides resources to communities in the county to improve infrastructure, attract industries and new jobs, remove blighted structures, and, in some instances, overcome damage resulting from natural disasters. Michelle works with neighborhood associations, civic and nonprofit organizations, and other stakeholders in the county who inform her about local problems and the aspirations of citizens. She also gathers information about the county's strengths and assets. The needs vary broadly, from 88-year-old John's neighborhood association submitting a grant application to improve safety on sidewalks leading to the schools; to Martha, whose two sons are MIA in two different parts of the world. Martha hopes Michelle can help her tap into resources provided by a veterans support organization.

Michelle brings long experience to her work. Before becoming director, she was the economic development coordinator who managed grants awarded through the county office to various cities or nonprofit agencies to improve the lives of people in a number of ways. Michelle is directly involved with recipients and intermediaries. Every day she receives several telephone calls and scores of e-mails from individuals across the county. She has established a high level of trust among the citizens and groups in the region. Her job is to assess needs fairly and allocate funds judiciously to improve the quality of life for as many people as possible with short- and long-term goals in mind. All spending must comply with the terms of the grant, and she must report accurately and regularly on the needs, costs, and outcomes of each project selected for funding.

Michelle's job is to learn about the communities directly from practice. Practice informs her of the veracity of the sources, the ramifications of action or inaction, the dependability of organizations or individuals who address problems, and the likelihood of positive outcomes. Michelle may be loved or resented at any given time depending on how constituents perceive her role. To some, she may be the admired community activist bringing benefits to the community. Others may look on her as intrusive, the bossy evaluator who polices the way communities use community development block grant funds.

Both David Farmer and Michelle Warden are public administrators dedicated to the cause of public service. Although both jobs are critical for a functioning democracy, they are fundamentally different in terms of interaction with citizens. For example, David has almost no contact with people outside his office and his social circle. He gets the information he will use each day from budget reports agencies send. Citizens are affected by services the state provides, but those citizens do not know David or his role in the process, and David is far removed from the citizens' response. David rarely has to leave his desk to get all the information needed to fulfill his job requirements. Feedback about how well he is doing comes from how smoothly the information systems operate and how closely various agencies comply with regulations.

On the other hand, Michelle works directly with citizens and monitors the pulse of the community. She knows the voices and faces of people affected by her decisions. She is informed from practice. How well informed she is depends on how well she understands her community and how openly citizens are able to communicate with her. Michelle can be sure people will call or write if a project goes wrong. She hopes to hear instead that an outcome is excellent. Information useful to Michelle is more about values than numbers, but because of the size of the grant and the diversity of demands, she finds herself spending considerable time recording and reporting numbers. Now the county has mandated all budgets must be submitted online via a secure portal, and the format for submission is driven by software Michelle is obliged to learn. She is increasingly frustrated as the digital age is slowly but surely taking her time away from her work on the ground. Michelle spends hours at her keyboard reading and reporting data, but software has no field or symbol to capture the values of John's neighborhood or Martha's nuanced needs while her sons are missing.

Michelle now spends more time processing information than she spends communicating with anyone in the community. However, the ease of transmitting data over the Internet generates a greater appetite for data. Soon all the grant winners will be required to submit data to support their progress through a federal data portal specifically designed to monitor and evaluate 501(c)(3) organizations. The same is true for Michelle, who must submit a quarterly summary data as a supplement to the cost and expenditure data submitted to the state accounting office where David works.

The state office has now set up a new Office of Grant Compliance and Monitoring (OGCM) to manage all data from federal grants about citizens who are being serviced through the grants. The OGCM will ensure HIPAA (Health Insurance Portability and Accountability Act) regulations and also mine data to evaluate which programs do better and predict how to develop new knowledge (technique) to reduce federal expenditures. Michelle's long experience and wisdom informed through practice with the people in her county matter less in her job performance. She subtracts hours she once spent in direct communication with stakeholders in her community because she needs time for administrative tasks now required by other administrators up the chain of command.

The above scenarios are real examples of how formal data gathering methods corroborated with information technology are eroding the value of public service. There are two implications we can draw from these scenarios. First, the overarching interest in reading too much from formative value-neutral data for monitoring and predicting service delivery does have moral consequences in a democracy. Second, the job associated with technology and technique receives wider attention in government thereby attracting a generation of students who are trained in *how* to fix a problem without the knowledge of *why* they are doing it.

Implications of the practice-administration dichotomy in the Information Age are clear. The Information Age has made it easier to collect and monitor information that can be captured digitally and transmitted through technology. The ability to proliferate documentation creates a thirst for more oversight, while the ease of archiving drives an appetite to track trends. Public administrators necessarily become cogs in the machine. They become unable to connect data with value to use information for the benefit of the human beings who need public services. Like a perpetual motion machine, the wheels of government deplete the energy they generate. The burden of a surfeit of data has even found its way into popular clichés. People interrupt with "cut to the chase" when they tire of distracting detail. "At the end of the day" precedes a brief summary. "TMI" is standard text to plug "too much information."

As governments perpetuate the insatiable thirst to collect data, public administration expends energy, time, and money engaged with machinery and numbers. Democracy loses when citizens become neglected bystanders on the edges of the bureaucracy they pay for and they need. This book is one step toward broadening our understanding of information technology and its scope for public administration in the Information Age.

Scope of Practice-Administration Dichotomy

The Information Age has presented us with new challenges. Technology has made it easier to collect and monitor the information necessary for the functions of a stable democracy. We have sophisticated tools to administer and monitor citizens, yet we fail to deliver tools that enable us to learn from citizens about their aspirations. The positive side of the Information Age is that technology has opened unprecedented opportunities for social mobilization by connecting individuals through social contact. The volume and breadth of information available has never been higher, and so administrators have at hand a viable asset for learning from practice. Information without context contains an intrinsic liability, because with raw information comes the challenge to interpret it accurately. Information with context is a resource for learning.

Despite renewed interest among scholars to focus on learning from practice, for example, from case studies and stories, the fundamental theory about practice-based learning has not received adequate attention. Whereas data-based learning informs us about the measurable hard facts about *how to* achieve certain goals, practice-based learning informs us *why* we need to achieve those goals. In other words, the hard data keep us on track to achieve certain ends given the means; the soft data create meaning for what we do, given our relationships. Today, with information technology being in the forefront, the distinctions between ends and means, values and tasks, are blurred.

The purpose of this book is to present the implications of the practice-administration dichotomy in the Information Age of the 21st century. The book devotes attention to the dichotomy of information as presented in figure 1 to show that the effect of this separation reverberates throughout the field of public administration. The principle of information has to be brought to the forefront of the public service debate. We must now question what we are learning and whether we are learning to serve or learning to adapt to self-evident outcomes based on the assumption that computing power produces intelligent behavior from an assemblage of formal rules, symbols, and codes. Throughout the book I discuss how to *inform* public administrators so they carry the *value* of public service to serve citizens as they continuously learn from practice, from contextualizing discrete data and codes to seamless operational information. As described in figure 2, the book is divided into two parts based on a conceptual understanding of information and the role of technology (part 1), and the operational philosophy of gathering value-based information for ethical leadership in public service (part 2). The first part is an exposition of the value of information and the value of public service as schematized in figure 1. Chapter 2 devotes the principles of information to explain the science behind the separation of formal information (data) from the informal (traces of data). Essentially this discussion sets the stage for the larger discussion on which type of data finds its best use for *informatizing* technology. The second chapter in part 1 (chapter 3) explains that information technology is the product of human action and a medium of human action. It is a social construct, therefore, whereas following Heidegger (1977), the chapter explains how technology reveals itself as an "ordering" mechanism of the social reality that we interpret through our own lenses. The co-creation and growth of humans and technology is fundamental to understanding the perspective that is argued regarding the broader goal of technology in society. The second part of the book discusses that given the Internet's presence in our daily lives, the informal information of the outer world is now within our reach. Chapter 4 lays out the foundation of information technology's utility for developing interpretive schemes to establish "relational lenses." Internet-based technologies can become interpretive aids that can enable us to contextualize known formulations to that of the unknown. By understanding the diversity of values, public administrators will be in a better position to *humanize* data and then *imaginize* pathways for solutions. These solutions are not only democratic but also have implications for building social relations and trust in government. Following the works of Anthony Giddens (1986, 1993), chapter 4 argues that context is the platform where "reflexive monitoring of conduct" takes place. Reflexive monitoring is the accountability of what one does based on what one is supposed

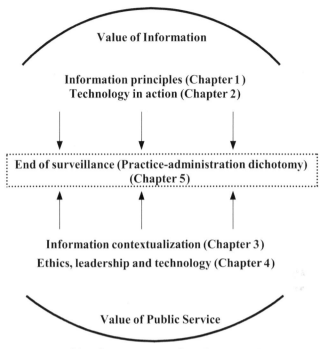

Figure 2. Public administration in the Information Age

to do. This argument sets the stage for a discussion on leadership and ethics of technology that is the topic of chapter 5. In this chapter we ask: How can technology be blamed when it is nothing but the product of our own desire? The question puts the burden on leadership and the institutions that are systematically being "battered" to fit the requirements of large-scale information systems. Philip Selznick's work has significantly influenced the arguments in this chapter. The chapter argues that organizations that remain technical instruments lack the distinctive competence and sense of an ideology to protect and conserve the values of the institution and the greater polity they serve. Following Selznick's appeal, the chapter notes that to build institutions, "creative men are needed . . . who know how to transform a neutral body of men into committed polity" (1957, p. 61). The chapter details the widespread use of data mining techniques and artificial intelligence to discover and predict citizen behaviors from formal data instead of learning from situations.

The final chapter (chapter 6) is a discussion on the implications of public administration in the Information Age. The final chapter shows why the value of information must be reconciled with the value of public service to take advantage of the emerging information technology in the Information Age.

The chapter explicates the overall significance of the practice-administration dichotomy in the Information Age by arguing that the control factor of technology has not only been dominant but also merits the accusation that technology is elitist and works against the values of the general citizens at the grass roots. The chapter calls for public administrators to take responsibility for the enormous citizen information at their fingertips. The challenge is to use that information and humanize it to understand values so as to know *why we serve* before we figure out *how to serve* or choose *who we serve*. This is the essence of public service in a technologically dominant global society. It is not surprising that many public administration scholars have also found technology at odds with democratic values when it is used (Kakabadse et al., 2003; Korac-Kakabadse et al., 2000; Haque, 2001, 2003, 2005). The chapter provides evidence to suggest we have been overly optimistic about the deterministic attributes of technology to measure and monitor programs for their success instead of learning from their failures. The "invisible forces of categories and standards" (Bowker & Star, 1999) are the primary ingredients by which society is reduced to data bits within an information system. Once the success and failure have been defined for the programs within the information system, the practitioners using the system consciously manage successes (what works) while undermining the feedback from practice-based knowledge about the lives that fail to meet the established standards.

The Information Age now presents us with a clear option of turning our lenses toward technology not as a panacea but as a vehicle for continuous learning from practice, a means to accelerate the realization of democratic ideals when administrators employ technologies to earn citizens' trust.

Part I
Value of Information

1

Introduction to the Theory of Information

Wherever there are messages, there is the possibility of information.
—Harold Garfinkel

Despite the agreement that technology is not a panacea and cannot solve the world problems, there is a certain level of confidence among us that technological determinism will somehow lead to a better human condition. This confidence is not only due to faith in technology but also information about others that can be used to understand, plan, and design better societies that give people better choices. We acquire information as data points to learn from them so as to fit predesigned models based on our conception of a better world. However, most of the time we fail to read information from our informal practices and interaction with others. Indeed, ICT has made us better informed technically, but socially, it has made us blind to our moral compass. This is because information that is technically appealing may not be the right choice to apply for socially and democratically desirable results.

The notion that information technology and technology in general can advance the human condition could be proven false unless we first address how it affects our values. We must find ways of nurturing those values that bind us to improving the human condition. Having information is known to be the necessary condition for making better choices, yet we falter in making decisions that are not only harmful to us but also to generations who follow such precedents. Due to ICT's technical advantage in processing and disseminating information faster than the human mind can comprehend, most information remains unused. Information overflow is now a common term that symbolizes information as a cheap commodity.

The value of information is demand driven. In other words, the value of information depends on how many people have used it or how many "hits" it has received from the cyber world. From the pile of information we find efficient ways to dig out what appears the most number of times. Frequency of usage determines value even if the information appears to be "dirt" (unethical) or even unlawful. Context is expurgated from the collected information, making the information purely neutral for universal application. Finally, what is learned from usage sets the moral tone of civil society. Indeed, the basis for information that could be valuable is lost in this process. In order to gain further insights on this issue, we must look into the principles of information in society.

Principles of Information

Information is the key to knowledge formation and the primary resource by which we can minimize our ignorance. The core philosophy of the classical Greek philosopher Socrates rests on the notion that "all that is evil comes out of ignorance." According to Socrates, the goal of the quest for knowledge is to remove our ignorance about how to manage practical life and the world around us. Drengson (1981) noted that it was Socrates' "awareness of ignorance, his own and others' that made him the wisest person in ancient Greece" (p. 237). We can avoid evil as long as we can use proper knowledge to minimize our ignorance. Certainly, according to Drengson, "it is the ignorance, as much as our knowledge, that defines or limits our world for us" (Drengson, 1981, p. 237). From this philosophical perspective we can argue that information in general is good because it helps us differentiate between right and wrong. If information has such qualitative value, how is it that although we live in an *infosphere*, we are unable to make the best of choices for improving the human condition? The answer is because not all information is the same.

The word *inform* stems from the Latin *informāre*, meaning "to shape or give form in the mind" (Oxford Dictionary, 2000; also see Latour, 2005, p. 223). It is an act of creation, to give form to something that did not exist. It can be seen as a value-added construct that can be further transformed as it undergoes translation from one individual or machine to another. For any form to be called information, it must shape the human mind. For example, the nature, the matter, and its usage that is still unknown to us has not revealed itself to us as information. The subject matters unknown to us are dormant information ready to be revealed once we can connect them to things we already know. The value-added portion of information helps us create the form—semantic or meaningful information that did not exist before. *Semantic information* encapsulates truth exactly similar to knowledge formation. Semantic *content* that has been falsified is *misinformation*. It is the factual semantic (meaningful) content that can be falsified so the truth is distorted. If the source of the misinformation is aware of the falsehood, then the information is called *disinformation* or distorted knowledge. If someone is misinformed and carries that information (without validation), they bear the burden of carrying false information, hence that knowledge has no value. However, individuals can be successful (in most cases they are) in using the false information in achieving their objective. The value of information is not dependent on the outcome but on the semantic information that encapsulates truth. Therefore, central to information is the ethical and moral dimension that carries the truth forward.

Information is fluid as opposed to discrete events. Once new information becomes available, it is *inscribed* and then mobilized through shared practices within a given network of association. This continuity of shared practice forms our minds, hence information. How and what is inscribed as the *new* information becomes fundamental to what information will become. No information can truly be passed on or shared because the inscription involves time, context, and understanding of what the information entails. Because what is passed on is only a part of the original that is transcribed, it becomes a matter of what is *not* translated (Garfinkel, 2008). When the information contains original truth, it retains its value without the vessel that originally held it. What is passed on to others becomes information for the next cycle, now ready to traverse on another vessel. The vessel, as I call it, is temporary as far as the information is concerned. The vessels embody the time, the context, and the people who hold the information. They are *traces of information* that can only be conceptualized if we understand or recreate the original past where the information was once inscribed. The vessel stays in the past but the essence of information is carried forward. Latour (1987) notes the "compromise between presence and absence is often called *information*. When you hold the piece of information you have the *form* of something without the thing itself" (p. 243; emphasis in the original). Therefore, we note that what is formed is the product of the human mind from what it gathers from the traces of information.

In modern terminology, information is often understood as data. Information and data are not the same. Information contains data but also includes the embedded meaning the data contains. The meaning cannot be expressed using data points because it is the vessel that holds the meaning and cannot be transferred over time and space. Floridi (2010) noted, "When we have all the data, but we do not know their meaning, hence we have no information yet" (p. 22). We can express the relationship between information and data as follows:

Information (I) = data (d) + meaning (m)

Data is the expression of the outcome, and the meaning explains what the outcome is. Let's say we have data for high school education and it shows 22%.

The meaning of what the 22% is—whether it is the percentage of people who are high school educated or dropouts—makes up all the value that this data will carry with it. Without the meaning, the data has no value. To get the information, we need to know the *meaning* of the data as much as we need to know the *data* (outcome) itself. If I learn (gather from practice) that the high school dropout rate is high, I do not know how high or whether it is high, without looking at the outcome (data)—I do not have the information, only the idea from practice of what it could be like (a lot of high school dropouts in a given area/situation). There will not be much value in incomplete information because it does not carry the weight of the *truth*. To *truly inform*, we can neither pass the vessel without the good, nor can we carry the good without the vessel (in which it was originally carried).

Information about people, time, and context carries the meaning of data. Essentially, the explanation of the data is embedded in the meaning of the data, which is the concomitant expression of the situation that can explain why the outcome occurred the way it did. It carries with it the stories that explain the *situation*. The *situational awareness* folds the people, time, and place of the outcome (data) under the *meaning* of information. Situational awareness is the embodiment of data. It can also be called the *traces* of information as discussed earlier. We can expand the meaning of data to include people, time, and place, which explains the situation:

$$Information = data + (meaning)$$

$$\underbrace{Information = data}_{Outcome} + \underbrace{(people + time + context)}_{Traces\ of\ information}$$

The mathematical theory of communication (MTC) founded by Claude Shannon (1916–2001) essentially deals with data without the meaning. According to Garfinkel (2008), "Information in Shannon's usage refers not so much to what one *does say* as to what one *could say*" (p. 103; emphasis added). MTC treats information as data with the primary purpose of devising efficient ways of encoding and transferring data. Information systems are developed to take advantage of this data. In MTC, the outcome of an event carries the weight of the information. In the high school dropout scenario, the concern for MTC is to deal with the number of high school dropouts as opposed to the overall meaning of that outcome reflected in the students who dropped out and the context under which the outcome was experienced. The outcome is quantified through some measure (20% drop out) or binary code (graduate = yes or no).

According to a thesis Shannon developed, the value of information can also be measured by quantifying how the information decreases the data deficit (that is, uncertainty). If we are certain about an outcome, the value of that information is low because we already know with certainty what the outcome is going to be (that is, there is no data deficit). However, when the outcome is not predictable, any clue that tells us something about the outcome becomes informative—it becomes valuable. In other words, if something is predictable, to inform us about it has almost zero value because we have not been notified about anything new (zero information). On the other hand, if we learn information about something that is unpredictable, it chips away our ignorance and informs us of something new. The relationship between information is inversely related to prediction. We can express the relationship as:

$$Information\ (data) = \frac{1}{predictable}$$

Therefore, in terms of MTC, when an outcome is less predictable, the value of that information increases. The increase in the predictability determines its value; therefore, the search for more precise data becomes an important goal because it makes that data more informative. Information is only a selection of the best possible symbol from a wide selection of symbols. In our example of high school dropouts, rather than expressing the outcome data for each individual separately, we can give the high school dropout rate expressed as a percentage of the total. One single number—22% (drop out)—becomes informative. The number tells us the type of school as well as the type of location the school is in. By reducing the *noise* associated with the data, we can make broad predictions about the school and its people based on the outcome data under investigation. Using a single number, we become *efficient* in predicting about the schools and localities in question.

According to the MTC perspective, if we consider using more data to explain the school dropout rate, we are in a sense compromising efficiency with more noise included in the data. Noise, in this instance, is the traces of information about the outcome, including the people, place, and many other attributes associated with the outcome. MTC studies information in terms of probability. The more probable an outcome the less information it carries. Information is essentially reduced to a statistical theory of correlation between the informer and the informee where the informer provides the causal clue to fit the needs of the informee's puzzle. MTC's central question is to investigate how much uninterpreted data (zero noise) can be transmitted efficiently. As Floridi (2010) notes, "MTC studies the codification and transmission of the of information by treating it as data keys, that is, the amount of detail in a signal or message or memory space necessary to saturate the informee's unsaturated information" (p. 45). Under the MTC, the data is a good representative to the fill the void, so it is considered well-informed data that provides the foundation for a mathematical approach to communication and processing. This is the study of information at the syntactic level. To make the data meaningful, however, we need to take the information beyond the syntactic level to the semantic level by including the traces of information as discussed earlier. To summarize: If we need to find out the quality of a school (thumbs up or down), we will be only interested in information at the syntactic level. If we want to investigate further about the truth as to why the quality of the school is good/bad, we would need to know the information's semantic content. Floridi (2010) compares syntactic information and semantic information to "the difference between a Newtonian description of the physical laws describing the dynamics of a tennis game and the description of the same game as a Wimbledon final by a commentator" (p. 48). Whereas syntactic information primarily deals with the scheme that leads to the outcome, semantic information gives life to the scheme by real-life description.

Classification of Information

The principles of information discussed so far give us critical insights to the nuances of information with respect to quantifiable information (data) and traces of information (people, time, and context). Information systems can be readily employed to transfer, replicate, and manipulate data. Since data is uninterpreted, it is free to be used in ways that serve the end users of the data. However, the traces of information are subject to interpretation because individuals trying to understand the traces must travel beyond their time and place to understand the contextual reality. However, the traces of information are not binary codes, hence not readily manipulable via information systems; human intervention is required to interpret the traces of information and "size" them to fit the information system requirements. To have a better understanding of the context, we classify the traces of information into different categories so we can organize our interpretation subject to the time and place and the context. Classification is an important methodological tool that has its roots in the Aristotelian tradition. The classification of information is a field of study that has received wide attention in anthropology, history, linguistics, physics, and other areas that deals with moral philosophy and metaphorical discourse. Once we classify information, we formalize the traces of information into a commodity that has consequences beyond what has been formed through interpretation within the formalism. As Geoffrey

Bowker and Susan Star (1999) note: "The moral questions arise when the categories of the powerful become the taken for granted; when policy decisions are layered into inaccessible technological structures; when one group's visibility comes at the expense of another's suffering" (p. 320).

We encounter all kinds of information on a daily basis that shape our minds. Human understanding of the natural and physical world depends on how we are able to simplify complexity by classifying items into groups. As we experience different social and physical phenomena, we first scan them to distinguish them from things we already know. Through initial scanning, we classify the items based on similarities with other phenomena already known to us. Classification of social and physical phenomena helps us simplify the real world to connect and make sense of the information derived from practice. In fact, according to Bowker and Star (1999), the act of classification is both organizational and informational and is always embedded in practice. Hence we scan all items that we encounter in our real-life experience and informally classify them in order to simplify our understanding of the world around us. However, consequent to our learning from practice, we start ascribing real formal meaning to our classifications, thereby creating a "black box" that carries its own meaning as it is passed on to others. Society suffers from "pluralistic ignorance" when it doesn't have access to the local practice, and instead has only the standard classification that defines that activity. Accordingly, the traces of information are now reduced to standard classes or codes. We are routinely conflating informal and formal information into our own vocabulary of classification. There is no correct way of classification, it is only the reflection of our own practice based on our moral and political beliefs that enable and constrain our actions to define who we are.

The information classification system is a science of its own. It is a formal field of study in anthropology and library science. Expertise also guides many classification systems. For every profession, there is some organized terminology used to describe particular *activities* related to performance and *outcomes* related to reporting. For example, medicine, nursing, botany, zoology, computer information systems, budgeting and financial information systems, coroners, machinists, law enforcement, and various professions have developed their own classification codes that guide their professional standards of practice. Although classification systems can help formalize action in terms of defining what the activity or outcome is, in reality all standardizations are inert, they have to be *performed* to *know* what the standardization does to real-life action. They must improvise and customize the action to fit individual needs before it can become habitual.

We can classify the social world into two broad classifications: one that is *formalized* and standardized to meet certain needs of society, and one that is *informal*, unorganized, unchartered, and open to individual interpretation. The formal world, which is already defined (given meaning), constrains and controls our interpretative lenses. The informal world is an unchartered territory yet to be revealed; we discover it through practice as we activate our interpretive lenses to form an opinion about it. If we use a framework that presents the social world in these two dichotomies, we can learn what controls and constrains our actions, and what enables our actions to be performed. In the following section, we will discuss the implications of the formal and informal dichotomy that shapes our decisions and actions.

Formal and Informal Knowledge

Following John Dewey's (1938) concept that knowledge arises from the active adaptation of the human experience, we argue that our experiences reveal to us the informal and formal dichotomy of the social world. We can extend our discussion on information to knowledge formation so we can argue that information helps shape our worldview through living and observing the acceptable norms and behavior of our society. This is called *informal knowledge*. Informal knowledge is the foundation on which we build knowledge about our society, its customs, norms, and behaviors (Rokeach, 1968). The conceptual understanding of the relationship between citizens and the polity is also part of the informal knowledge. It yields the primary cues about what is right and what is wrongly demonstrated in the norms prevalent around us. Informal knowledge explains the qualitative aspect of the social world for which there are no measurement tools except for society's moral and political dimensions. This is the sociological interpretation of knowledge formation that forms the basis of values (Rokeach, 1973).

Formal information is quantitative in nature and has the advantage that individuals can weigh that information with respect to whether the information is beneficial or has some strategic value to get things done. This type of information helps us build *formal knowledge* or *strategic knowledge*. We find it necessary to gather this type of knowledge because we find it useful to improve our human condition and, in many cases, useful for our survival in this increasingly complex and competitive society. This is the economic rationale of knowledge formation (Sowell, 1980; Hayek, 1983). Individuals are motivated to gather this type of knowledge for self-interest or to gain power and prestige in society. Where informal knowledge shapes one's comprehension of society, formal knowledge builds on informal knowledge and signals strategies to gain future advantage. Both types of knowledge shape us as humans; they are the roots of the knowledge we form and the actions we take as a result.

In a lifetime we gather both types of knowledge. The bias toward one type of knowledge over the other may be cultural or dependent on individual needs that vary across societies. The formal knowledge carries more weight in society because society deems it necessary that such knowledge is required for individual and societal growth. Formal education would be a good example of formal knowledge, followed by a variety of technical and expert knowledge that gives legitimacy to such a knowledge base. The type of knowledge dominant in a person's worldview influences that person's actions. A proper balance of formal and informal knowledge is of critical importance to maintain a balanced society. The diagram below shows the continuum of a formal and information knowledge space. The horizontal axis shows informal knowledge (I) and the vertical axis shows formal knowledge (F).

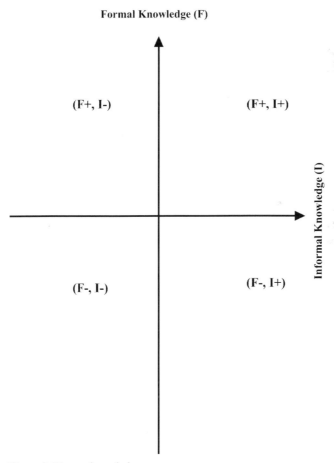

Figure 3. Human knowledge space

Given that information leads to knowledge formation, we can argue that the individual whose formal knowledge is dominant (F+) compared to informal knowledge (quadrant I and II in the diagram) is likely to use the knowledge to compete with others in order to improve his or her own human condition. In modern-day industrialized economies we see such examples of bias toward formal knowledge to gain the competitive edge. Modern societies are more data driven, and there is inherent bias toward formal knowledge that is for evaluating and monitoring individuals and their associations.

On the other hand, in those for whom informal knowledge is dominant (I+), greater is the influence of the social norms that bind a society together, and greater is the inclination to use that knowledge to preserve those values (quadrant I and IV in the diagram). Informal knowledge is expressed through human relations, and trust plays an important role in our faith whether to invest more or less of informal knowledge. In less developed countries, individuals tend to focus more on informal knowledge—family, ethnic, or cultural ties—to advance in society. Informal knowledge dominance in lesser-developed countries has greatly affected their democratic establishments, including public institutions.

Formal and informal knowledge are relative to each society and can only be compared by the relative intensity of one type of knowledge over the other. Although the diagram is a simplistic representation of how our knowledge dominance affects us, it is one way we can analyze the individual knowledge continuum that affects individual actions in a given society.

The Theory of Information Flow

The flow of information is intrinsic to how knowledge can influence society. It also has implications for formal and informal knowledge gathering. The Internet revolution has shifted the information landscape toward instant information access and ubiquitous knowledge formation beyond political and cultural boundaries. A large part of the process of gathering information depends on how the information is able to mobilize large groups of people in a short amount of time. When information is very useful to others, the likelihood of speedy dissemination is greater than when the information is less useful. In other words, the intensity of information flow is proportional to its usefulness. As more and more people find relevance in the information and find it can have positive impact on their individual goals and outcomes, the flow of information is likely to increase. It is important to note that if someone conceals a part of the information because of its strategic value (formal information that gives them a competitive advantage over others), the flow may slow or stop. Information carrying strategic value is not considered everyday information that has everyday value to common people. They may emanate from informal information to form formal knowledge through practice. Once a group of users realizes the value of this informal knowledge in practice they may restrict the flow of information to protect their common interest. For example, if I find riding a bike to work is easier and cheaper than driving (given I don't have to spend time to find a parking place and spend money to pay for it) I may share my new venture (information) with a neighbor or colleague, who may then share this with his wife, who then shares it with her colleagues and others. As long as people find their own interests unaffected by sharing the information, the information is expected to flow unrestricted. But what if adoption of the new venture means spaces become scarce at the bike rack? Sharing the information causes the costs and risks of inconvenience to increase, so the marginal costs of sharing outweigh the marginal benefit of sharing. We can see in this analogy that information shared without disadvantage to a large number of people is likely to spread faster and more broadly than information that carries strategic value. We must also note that the assumption that people are unlikely to share information that carries strategic significance is based on a societal value—self-interest. The value of self-interest is self-imposed and has been criticized on the grounds that individuals are willing to cooperate and share information based on reciprocity—if others in the community also abide by similar notions of sharing. As noted by Elinor Ostrom (2000), the "world contains multiple types of individuals, some more willing than others to initiate reciprocity to achieve the benefits of collective action" (p. 138). Ostrom argues in a society with a high degree of trust among individuals, a "rational egoist" is likely give up their ego and adapt to behaviors that increase socially desired "common pool" resources. Therefore, in a rational world where people are motivated by self-interest, the flow of valuable information (perceived value) will travel slower than the flow of informal information. However, as noted here, formal information that is perceived to carry higher value can also flow faster if there is a higher level of trust among the information users.

In primitive societies, the role of formal information is very limited, so, consequently, the level of trust among tribal members is very high. In a hypothetical world, if all individuals treat most of their information as informal, we can expect the individual members to share a high level of trust. Whenever information is formalized, in other words, standardized with formal codes, it is stripped of its original garb of neutral-colored clothes. Its originality is hidden; only those who have used the information in practice can realize its true meaning. Even with the best of information, a society may become knowledge stagnant—to learn any mundane or routine subject there would be people formalizing the information to provide expert knowledge. Therefore, even informal information can incrementally coalesce in the hands of small groups. There are important implications of power and dominance when such an informal world is formalized. Citizens who pull themselves from the practice world and interact off the ground are more likely to be influenced by formal knowledge. The surge of interest in online apps for different services is one sign of such formalization of knowledge.

It is not surprising why information that is new (not common knowledge) is privately owned and thus more valuable than the information that is available publicly. Public information was not as valuable until recently, when technology allowed parsing out millions of publicly available data to tease out strategic information for private use. Such data mining techniques have become one of the most popular ways of "cashing out" from mines available throughout the public landscape. The rise of individualism, the radical disengagement from social life, and the polarization of income classes is vivid evidence of the growing interest in formal information as opposed to informal information in the lives of ordinary citizens (see Bellah, 2008).

Having more information does not necessarily lead to an increase in the practical knowledge base. For instance, information technology, particularly the Internet, has reduced the cost of information and led to a surge of information that has no parallel in the history of our civilization. Despite the unprecedented flow of abundant information, we see more uncertainty than before, with an uncontrollable stock market and the collapse of economies from Europe to the Americas. There is an apparent disconnect between what information is accessible and what is used to improve the human condition. Although we know information is valuable, we are unable to select what information will be useful at a given time. Information has little value if it is not used, yet the abundant volume of data accessible does not guarantee more information will be used. Therefore, information flow is a necessary, but not sufficient, condition for information usability for improving the human condition.

Formal Information and Public Policy Making

Because of our limited capacity to process information, information technology becomes handy in helping us store, monitor, and process the information. As noted, information technology is more receptive to formal information (data) because it can be quantified and measured. By its sheer nature, information technology is inadequate when it comes to information that is subjective and interpersonal in nature.

The limitations in the ability of information technology to process both quantitative and qualitative information have important implications for public administration and social science research in general. It is not surprising that information technology's ubiquitous presence as the primary tool for making policy decisions has affected the type of policy decisions that take priority in government. For example, elected leaders favor policy decisions that are objective and scientific, making them more likely to fund academic disciplines that favor scientific methods (Cohen, 2009). Norma Riccucci (2010) argues the tension between qualitative research and quantitative research in public administration is real, to the point that often qualitative research is considered less rigorous ("soft science"), nonscientific, and hence less useful (p. 42). Today quantitative and statistical techniques take over the real world problems we face. Quoting Joseph Nye, Riccucci argues, "the motivation to be precise has overtaken the impulse to be relevant" (Riccucci, 2010, p. 43/Cohen, 2009, C7). Accuracy of knowledge trumps the ego's desire for relevancy in today's technological sphere. It is more important for information to be correct. Delivering information with precision is where the information carrier gains relevancy.

Needless to say, because of its processing power, information technology is the primary vehicle by which scientific techniques gain supremacy. Numerous practical applications can be formalized into techniques that are readily available to cause an effect through the push of a button. In the academic world, in public administration, and in public organizations, the type of information used matters greatly. In a democracy where technology is being used in diverse aspects of management—from street cleaning to disaster management to providing basic human needs including reducing hunger and eradicating diseases—we must seek answers to the question: What information technology is best suited for using informal information to increase citizens' trust? Why do we have to accept a blanket application of information technology in affairs of government? The question is not whether we have to adapt to technology; that is obvious. The question we must ask is—adapt to what type of technology and why? After all, as Todd R. La Porte stated about technologies in general: "There is nothing inherent in the structure of these technologies, the physical and biological laws upon which they are based, or the institutions associated with them, that should leave us awestruck and submissive in the face of them" (La Porte, 1971, p. 70). Indeed, the fact that technology is a social construct has implications for how society shapes technology and how technology mediates within society.

Technology cannot be seen as independent of human social values. If its use is to improve the human condition, it must serve the larger social order congruent with public values. It is the institution's task to use technologies that will enable positive values to be preserved and cherished. More bluntly, informal information must take the same priority as formal information, regardless of which data can be quantified and deemed acceptable to technological demands. This will require public institutions to develop a refined sense of social purpose geared toward increasing information flow and access to a practical knowledge base. Information technology in this regard should be used in processing information that translates social values. Technology thus can become a tool to increase public trust in government.

Conclusion

There is an apparent disconnect between government and the citizens it serves. To fill the gap, information has been identified as one of the "basic and necessary bridges for transforming governance in the 21st century" (Kettl, 2002, p. 169). The ubiquitous presence of the Internet in our lives tells us that the technology bridge is strong and growing. However, that bridge is only getting longer and longer as information technology is institutionalized with quantifiable formal information taking precedence over informal information, while citizens and their values recede from influence on government. Public managers must heed the calls of social entrepreneurs and civil society organizations to use information technology to address basic societal needs. This requires a closer look at public institutions and their dependency on rational instruments to solve social problems.

2

Information Technology in Action

> When we once open ourselves expressly to the *essence* of technology, we
> find ourselves unexpectedly taken into a freeing claim.
>
> —Martin Heidegger

Information technology is a product of human action as well as a medium of
human action (Orlikowski, 1991; Orlikowski & Robey, 1991). It is recognized
as integral to social development as it mediates into practice. When humans
engage in productive activities, they seek technological help only to routinize or "carry down" the *idea* in order to continue the activity. Following the
work of Martin Heidegger (1977), the tools and the social world can be seen
as "standing reserves" waiting to be revealed by someone whose imagination meets the challenge to use the reserves for advancing their *ideas* of the
productive activity. According to Heidegger (1977), "the challenging gathers man into ordering" (p. 19). With human creative endeavor, technology
comes to aid and formalize the process, which we often refer to as the *technique*. The technique is the formative nature of a process that describes human labor, motivation, and cognition organized toward a desired result. Following this concept, technique is the *art of the mind—poiesis*—which is given
its formative shape through technology. Technology reveals itself as the formative answer to operationalize the technique. Hence, technology is often
seen as an independent external force to influence the nature of a given outcome. Using Heidegger's interpretive lenses, we can see how technology is
expected to have a deterministic impact on human behavior and organizational outcomes. However, note that the art of the mind is the primary initia-
tor of the formalism that is taking place through technology. Therefore, in
a progressive society, technologies are central to human and organizational
development, and human organizations are central to the development of
new and emerging technologies. The co-creation and growth of humans and
technology is fundamental to understanding the perspective that is argued
here regarding the broader goal of technology in society.

When technology becomes routine and integral to an organization's performance, it becomes an active part of the social life of the organization. Like any physical object, technology in itself is inert and passive until it becomes an active resource for human use. Technologies are a product of their times and therefore are socially constructed. Both technologies and institutions undergo dynamic changes in terms of how individuals and interacting communities within organizations use them. Technology and society are not independent spheres. Rather, they influence each other; they are mutually constitutive. Therefore, an informed perspective on technology is critical to understand the relationship between human and technology and how this affects practice.

In this chapter, I explore the theoretical underpinnings of technology, particularly its claim to transform society and its limitations in doing so. Essentially, the battle is to overcome the control-oriented attributes of any technology that has become the effective instrument for maintaining and reinforcing "disciplinary power" (Foucault, 1977) in a technocratic democracy. The focus will not be to inculcate a pessimistic stance, but rather to encourage deeper insight to explore relationships and guide a generation of public administrators who have already embraced the new information technology revolution.

The Human Experience and Technology

Technology can be viewed as a tangible physical object such as hardware, software, or a machine that humans make productive use of in order to assimilate and process goods and services. The ontological assumption of social reality as subjective or objective has often sparked a debate among social scientists about the role of technology in society. The *objective* camp has long held the deterministic role of technology, arguing that technology is an independent object that can have direct impact on organizations, thereby constraining human action (Ellul, 1964; Marcuse, 1964). This dystopian notion has powerful implications for how technology and institutions dominate humans, and how the artificial world takes over the natural world and reduces it to virtual (false) reality. A deterministic view became more popular during the early computer generation and the last couple of decades. Technology is not only the means to an end; it also becomes the end in itself as technology earns the right to rule, as humans wait at the receiving end of the technological onslaught. In contrast to the dystopian view of technology, the more recent *subjective* camp sees technology and globalization as forces that can liberate society and solve most of societal problems, including poverty, crime, disease and even personal problems. The optimism is shared among many technology enthusiasts and policy planners who view technology as a possible vehicle to solve many societal problems. The idea of the technological fix is common to many popular beliefs about technology because artifacts are generally seen to instill rational behavior. Therefore, in this subjectivist view, society is in the driving seat to influence technological prowess to empower and liberate the marginal class. Artificial intelligence has gained prominence under this utopian view, which claims one day computers will be able to replace humans as they will understand, see, and learn from human behavior (Minsky, 1986, 1985; Minsky & Papert, 1973). The limitations are not within technology, rather in our cognitive capacity, purposeful planning, and institutional barriers that prevent us from realizing many social goals. Both of these views have their merits for understanding the role of technology in society. However, technology does not operate in a vacuum; when it is created there are many assumptions made about how it will be used, who will use it, and where it will be used. Once it is made, no assumptions hold about how it will be used and how it will influence individuals to create an impact in the larger society. The philosophy of technology is as wide as our imagination.

Explaining the philosophy of technology, Don Ihde, in his important work *Technology and the Lifeworld* (1990), notes it is critical to take into consideration the interpretation of the human experience in relation to technology to explain technology's role in society. Ihde argues that to explain the role of technology, the experience itself must be understood in its totality to be constructive and enduring. "Such a science," he states, "cannot simply reduce its field to some arbitrary aspect of the whole. Both 'external' and 'internal,' 'subjective' and 'objective' aspects must be included" (p. 23). He suggests it is critical that we employ a "non-reductive strategy" with respect to the field of inquiry (p. 23). We must also define the lens we use to interpret the human-technology experience because our definition essentially sets the stage for what we are observing—the ontological status of being. Don Ihde notes that interpretive strategies are misunderstood as a purely subjective analysis, particularly in explaining the philosophy of technology, because we ourselves are attempting to interpret our relationship to technology without clarifying our relative status to the material world. We have failed to explain the inquiry as a relationship where the object being studied and the individual studying it are in a relativistic position, rather than in a privileged position. The arbitrariness of assigning superiority to technology over humans, or vice versa, sabotages the relationship from its inception and prejudices its outcome. The privileged status given to one over the other has in fact created diverging and often extreme views about technology's role in society—from deterministic, nondeterministic, utopian, to dystopian views of technology.

The apparent disconnect between the broader social world and the material world is embedded within our dominant worldview about facts and values. Scholars dealing with technology can often be distinguished based on their normative worldview about the role of objective facts, including tools and subjective values (social values and politics) that transform our society.

Technological Determinism

The most commonly held view regarding technology is that it is neutral; technology is an instrument and primary driver behind changes toward a desired outcome. Therefore, given the availability of the "right" technology, substantive social and political changes can transform societies, even civilizations. The deterministic assumption follows that the total responsibility of using technology lies with the user (a "guns don't kill" argument). It is only a tool, and its application is only contingent on societal values. Therefore, unlike institutions that cannot be readily transferred from one society to the other, technology is universal, noncontextual, and transferrable barring only the cost. This utopian view holds that the same standard of measurement can be applied in different settings, leading to universal efficiency gains regardless of culture, time, and even civilization. The proponents of the instrumental view of technology envision a world that can be perfected, both in design and outcome, through technological domination. This instrumental view of technology is prominent in mainstream social science literature, including the literature of public administration.[1] The mainstream instrumental approach to technology has been criticized by a growing number of scholars best known through the writings of the French philosopher Jacques Ellul (1912–1994) and the German philosopher Martin Heidegger (1889–1976). According to their work (also known as the substantive theory of technology), technology is constitutive because it restructures the social world through technical rationality. Technique is autonomous because it becomes a symbol for development and efficiency. Its self-enforced rule is counterproductive to human development because it drives "helpless" humanity toward an automated and self-regulated destiny that can have unpredictable consequences in social and political life. Ellul and Heidegger reject the instrumental view on the grounds that it devalues humans as mere objects consumed by desire to feed their own self-interest. Technology appears as the only means to serve that purpose. Humans use technology in search of quick fixes and efficient techniques to meet the ever-increasing demand of individuals and the collectivity. With this concomitant phenomenon becoming a value in itself, I have summarized it in the next section.

Sociology of Technique

Heidegger (1977) explains that in order to understand the role of technology, we must understand what *causes* technology as described in Aristotle's ontology. Technology is the formative tool that holds technique, and the technologist is the organizer who plans how causation will be executed. Cause, he explains, has four categories: (1) *Causa materialis*: the material that is required to initiate a cause. For example, we need brick or wood to bring about a house. Similarly, we need *information* in order to inform someone about a fact. (2) *Causa formalis*: "the form or the shape into which the material enters" (p. 6). In our example, the bricks and wood need to be of certain size and the information has to be "shaped" into data (noises removed) in order to pass on the *targeted* message. (3) *Causa finalis*: The process or activity by which the material is transformed into a new product. Here again, the bricks have to be aligned, and mortar needs to be applied methodically to bring about the house. With information, it has to be processed or formatted according to the desired result. (4) *Causa efficiens* is what brings about the action or effect of the actual finished product. For example, the builder or the craftsman causes the effect to take place. The information system or database manager plans and organizes what information should guide the action. Heidegger argues that although we may generally focus on the objective result (*causa finalis*), or the final product that is produced, we must also note the importance of the principal architects who are *responsible* to formally align the other three causes toward a goal. What then is technique? According to Heidegger, it is the four causes that make the technique or *instrument* that "brings-forth" or "reveals" what was once "concealed." He notes technique is the mode of *aletheuein* (getting to the truth). He elaborates: "It [technè] reveals whatever does not bring itself forth and does not yet lie here before us, whatever can look and turn out now one way and now another. Whoever builds a house or a ship or forges a sacrificial chalice reveals what is to be brought forth, according to the perspectives of the four modes of occasioning" (p. 13).

So the essence of technique is reflected in technology through the causal process that is embedded in the object. In other words, as Heidegger noted, "instrumentality is considered the fundamental characteristic of technology" (p. 12). Technology is not the means, but a way of revealing what was once unknown. It is a process by which we assimilate and organize a system to reveal the *truth* that was unknown to us (p. 13). Technology, therefore, is not merely a machine, but an organized *enframing* of the human mind—"it is the way in which the real reveals itself as standing-reserve" (p. 23). Therefore, at a given point in time information technology folds the information and the traces of information into an organized artifact.

In order to understand the role of technology in human affairs, it is critical to further explain the interpretive analogy Heidegger described. Technology is not an instrument to get things done. Rather, it is an instrument to reveal the truth—to bring forth new lenses so we can focus on things that were not apparent to us before. In this sense, technology has a moral dimension. The essence of technology is to organize and assimilate the causes that explain and show new pathways to improve the human condition. This perspective is critical to put technology in its right place so that we do not succumb to use technology for self-interest (means to an end), but rather as a tool to organize and assimilate our ideas from practice to understand others (values) and to improve the human condition. Because the ends are unknown, the proper use of technology should reveal what is not apparent and let the decision evolve out of the human experience.

Jacques Ellul has eloquently argued the dangers of having faith in technology to develop techniques to meet the demands of a self-serving polity. He argues that the appeals of technological power are objectivity, mathematical precision, and scientific understanding of the world both seen and unseen. According to Ellul (1964), the focus of the objectivist-scientist is not so much on technology, but on the *technique* that can produce more of the "good" with the least possible cost. Technology is only the *means* to develop a *technique* to meet the individual consumer demand. Therefore, according to Ellul, "science has become the instrument of technique" (p. 10). He ponders that scientists might act more prudently when using their own discovery and innovation, and might even be afraid to launch their carefully calculated laboratory findings to the world. But, he asserts, "how can he resist the pressure of the facts? How can he resist the pressure of the money? How is he to resist success, publicity, public acclaim?" (p. 10). The scientist is run over by the technologies' linear logic despite the logic being at odds with tradition and history. Ellul's skepticism is rooted in a trust in tradition and human nature akin to the philosophical tradition of Edmund Burke, the 17th century English philosopher.[2] Ellul argues, "Technique has become autonomous," and it "no longer rests on tradition, but rather previous technical procedures; and its evolution is too rapid, too upsetting to integrate the older tradition" (p. 14). To replace tradition by procedure is to reduce collective social values into techniques for solving social problems. Knowledge that is fundamentally rooted within traditions (see Sowell, 1980), takes the back seat to techniques.

Jacques Ellul is not opposed to science or technology. To the contrary, he argues that the instrumental view can do more harm than good to society because instead of using collective social values to drive technology, we may be tempted to use technology for better "techniques" to feed our self-interest, even if the cost is borne by others who do not benefit. The competition for technique is due to the overriding need to have competitive advantage for more wealth.[3] It is the competition for technique rather than technology that is the fundamental social concern.

Technological Rationality to Political Rationality

The rationality argument that is embedded within the technological environment pervades political rationality. Marcuse (1964) states that the "technological rationality has become political rationality" (pp. xv–xvi). He explains that the larger implication of technology is not primarily limited to technology, but to faith in technological rationality that directly influences political behavior that is then well reflected in technological design.[4] This may have been the ultimate fear of Ellul and Heidegger, whose attempt was to extend the idea that the narrow technological view expressed in terms of technique is unable to address the plurality. Andrew Feenberg (1991) clearly makes this linkage through his *Critical Theory of Technology*: "The values and interests of the ruling classes and elites are installed in the very design of rational procedures and machines even before these are assigned a goal. The dominant form of technological rationality is neither an ideology (an essentially discursive expression of class self-interest) nor is it a neutral requirement determined by the 'nature' of technique. Rather, it stands at the intersection between ideology and technique where the two come together to control human beings and resources in conformity with what I will call 'technical codes'" (Feenberg, 1991, p. 14). What is profound in Feenberg's argument is that political rationality, once formalized into a technical system, can generate autonomous responses as per design. Once a political decision is codified into a technological system, no matter how small the inefficiency is, it can produce mass scale inefficiency. The political intrusion into technological artifact raises Orwellian concerns in a democratic society. The implication of Feenberg's argument should carry great weight for contemporary public administration circles, at least for those who see a broader role for technology in decision making. For example, the widespread interest in the so-called police technologies, such as using voice recognition to scan mobile networks, tracking a citizen's every movement using GPS and RFID technology, and changing e-mail contents while en route to a recipient. Recent interest in drone technology to monitor traffic movement and emergency evacuation can be extended for purposes that violate citizens' privacy concerns. Some technologies are also available to secretly turn on webcams built into personal laptops and microphones in cell phones not being used. Clearly, if an ideology that goes against any democratic norms can be technically rationalized, it can turn a democratic state into a police state.

Technology today is capable of transferring efficient systems across the globe almost instantaneously. Given the discussion about technique and technology so far, we can make some important assumptions regarding the implication of technology in a digital global society. If technique can meet four basic requirements, it can have a significant impact on the digital world. First, the new technique must have an ideological foundation. Second, the technique has to be codified into a technological system so it can be packaged as a holistic system that is transferrable. And third, the new technique's performance can be measured through quantifiable objective criteria. Finally, the measurable outcome is subject to "customization" (manipulation) based on the objective criteria of the owners. This instrumental view is dominant across governments in the Western world and is widely seen as a norm across the globe. In addition to the dangers pointed out earlier, in a democratic society there are fundamental flaws inherent in the instrumental view that justifies political rationality. First, the technique is noncontextual; it assumes the neutrality of time and place. Second, the outcome is developed without input from people who will be affected by the application of the technique. Finally, the system is primarily accountable through measurable objectives. The limitations clearly highlight how technical rationality used for developing "public instruments" for public purposes violates basic democratic practice. Such public investments can be a great burden for democratic institutions, not only in terms of costs to society but also in choices with far-reaching consequences for citizens. As noted earlier, the way technology is embedded in institutional daily routines critically affects the way public values are nurtured in a democracy.

Instrumental Rationality and Failures of Public Technology

It is not surprising that public sector technology is greatly influenced by the instrumental view of technology because almost all technological systems (defined here as group techniques) are produced in mass scale by the private sector with private sector objectives in mind. As Bozeman and Bretschneider (1986) noted, "design, implementation, and evaluation of management information systems differ between public and private sector organization" (p. 485). The authors expressed skepticism about using management information system (MIS) in public agencies because it is primarily used in the private sector as a "means of exercising and augmenting control" (p. 484), which is not compatible with democratic, public institutions. In addition, they argued that decisions to procure large-scale technology investments are generally made without direct input from the line workers who are responsible for data input and routine management of the technology systems. Decision makers who are involved in technology investment decisions include the heads of departments from the larger unit of governance, the central information technology department, and the finance department. The exclusion of the primary end users of the technology is counterintuitive. In another, more recent work, Bretschneider echoed the same sentiment, adding that because the diffusion of technology is largely driven by end users who try to solve specific problems using technology they, not the technical elites, should be considered the primary stakeholders in decisions to procure technology (Bretschneider & Wittmer, 1993). From initial design to procurement, the process follows a rational decision protocol that is antithetical to public institutional values. Kraemer and Dedrick (1997) argued, considering the broader implication of technology in government, that constitutional issues can be raised particularly in regard to congressional oversight of public technology.

The increasing infusion of private sector values in public sector technology management creates disincentives for technology adoption and diffusion in public agency. It is not surprising why resistance to adopting large-scale MIS within public institutions is so high. Maureen Brown (2003) cites a series of studies to highlight the public sector technology adoption crisis. She notes:

> In their attempt to "seek consonance in information systems" Klein and Jiang (2001) claim that the "failure rates for system development projects are alarmingly high" (p. 195). In a 1998 survey, Gallagher (1998) found a 76% failure rate in enterprise management solutions, and Ambler's (1999) study of system development efforts claimed that 85% of all projects end in failure. Given the difficulties associated with the latter stages of e-government as well as the high failure rates of technology in general, the Gartner Group (2000) predicts failure rates of 50% or more for e-government. Calling attention to the "software crisis," Zmud (1998) asserts that software projects typically go over budget, go over schedule, and fail to adequately perform tasks they were intended to conduct. (Brown, 2003, p. 346)

If an individual or private entity were paying for this sunken cost, they would have been bankrupt multiple times. Failure here is not so much that technology is not useful. The failure is in the institutional process that tries to inject technology into administrative management. Brown's research demonstrated stakeholder involvement is critical for the successful adoption of e-government projects. By conducting telephone interviews with 28 state chief information officers (CIOs), Brown found 82% of the CIOs indicated stakeholder involvement had at least a medium impact on success, and about 40% mentioned stakeholder involvement had a high impact on "curtailing problems" (351–52). Stakeholder involvement is not only compatible with public institutional values, but is also a commonsense approach in view of the fact that successful implementation cuts waste and increases citizen trust in public institutions.

Instrumental Rationality and Private Sector Influence

Stakeholder involvement within public institutions is an indirect approach to citizen-driven information management application in a democracy. While it is sometimes impractical for citizens to get directly involved in public technology investment decisions, street-level bureaucrats can be the best representatives to recommend the best tool to address the problem of the day. They can also check the abuse of power and waste becoming common in the procurement and application of large-scale information management projects. An elitist, expert-based approach, even with congressional oversight, is a narrow instrumental view that only rationalizes the *use of the means* to meet their *defined ends*. Indeed, as information technology and technological systems become acknowledged as substantive players (already there, though underrecognized) in the administrative process, the street-level bureaucrat's role will be reduced to that of a tool to *manage* and *maintain* the technological system. The discretionary authority of the street-level bureaucrats will also significantly diminish in the technological era. As noted by Bovens and Zouridis (2002), we are now moving away from street-level bureaucracy to screen-level bureaucracy and presently to system-level bureaucracy. The "expert systems," they argue, "have replaced professional workers" to the extent that "the process of issuing decisions is carried out—virtually from beginning to end—by computer systems" (p. 180). From database managers to system designers, system analysts, public information officers, and legal compliance officers are running the public bureaucracy system. Bovens and Zouridis note that "the hundreds of individual case managers have all vanished. Their pivotal role in the organization has been taken by systems and process designers" (p. 180). It is alarming that public administration education is not prepared to supply students with the skills needed to manage system-level bureaucracy. Business schools and CIS programs are filling the void to meet the management career demands in government today.

Almost all decisions about technology are conducted via private vendors and contractors. In addition to the private sector values infused into technological procurement and management, almost all technology is designed to operate purely in the private sector. For example, all federal agencies are now under the enterprise architectures (EA) system that operates fundamentally under a business plan that connects each federal agency under one integrated system. The business plan of the US government, released by the Office of Management and Budget (OMB) in August 2011 under the title *Capital Programming Guide*, highlights that the purpose of the guide is to "provide professionals in the Federal Government guidance for a disciplined capital programming process, as well as *techniques* for planning and budgeting, acquisition, and management and disposition of capital assets" (OMB2011, p. 1, emphasis added). OMB expects that the integration of federal agencies will not only cut waste by identifying duplication but will also systematize the governmental mechanism through monitoring and collaboration. The implication for organizing the federal government's financial system under the grand enterprise architectures scheme is to connect all federal agencies and their programs under one monitoring tool.[5] This goal is clearly reflected in the federal government's popular IT Dashboard: www.itdashboard.gov. The purpose of the Dashboard is to allow citizens and federal agencies to view the costs of federal projects and agency performance and to rate them based on objective quantifiable data. This transparency is a welcome note for citizens who can now have access to data previously unavailable. However, transparency through numbers (and rating of projects and agencies) may be of interest to federal government and media more than to private citizens. This is especially true when citizens like to see the impact of federal dollars expressed in terms of social investments. Many states are also moving toward replacing enterprise resource planning (ERP) systems with enterprise architectures, which integrate a variety of business information such as accounting and human resources data across governments.

An important study by Marie-Claude Boudreau and Daniel Robey (2005) found "as technological artifacts become more tightly integrated into larger systems or networks, a narrower range of enactment may be expected from users" (p. 5). This study follows Wanda Orlikowski's (2000) findings that "integrated systems not only make individuals more dependent on the systems that are inflexible," but also "reduce the degrees of freedom available to users" (p. 424). Clearly, the drive for better techniques for managing government is no longer debatable. The implications are significant. By taking away the human discretionary authority, bureaucratic experience from citizen interaction cannot be informed or applied into practice. Public values can now essentially be replaced by techniques.

Another important aspect of private sector norms influencing government technology investment is commercial software. Commercial software allows flexibility in terms of implementing technology to the specific needs of organizations. It can also increase the interoperability to meet federal, state, and local report mandates. Recently, the Chief Financial Officers Report on 20 years of the Chief Financial Officer Act presented to Congress noted, "Today, the Federal government is almost exclusively using commercial software adapting its processes to tried-and-true practices rather than creating unique software that would be unable to communicate with software in other systems. The opposite was the case 20 years ago prior to enactment of the CFO Act, with individual agencies most often developing their own financial information technology software" (Taylor & Rymer, 2011, p. 21). Governments are exclusively dependent on private sector techniques to make the task of government efficient and targeted to meet predesigned objectives. One implication of the commercialization of federal systems is that commercial software is tailored to business needs. Do all agencies operate as a business enterprise? Does it serve the public purpose if they are trying to optimize their return on investment (ROI)? Technologies are appropriate to meet agency goals when agency targets can be objectively quantified. Otherwise, technology would be more of a symbol of power and legitimacy than a tool that helps solve substantive problems for the agency.

Lawrence Lessig (2011), in his pioneering work, argues that the republic is lost because we are obsessed with individual rationalism. The financial meltdown of the century, the mortgage meltdown, and the subsequent Wall Street collapse, he argues, was not due to some kind of craziness. The core driver for the meltdown was individual rationality, which in collectivity was irrational and crazy. Technology was the enabler that allowed people to use computing power to gamble with the system and concoct complicated "financial innovations." This so-called systematic rationality spread to the masses through computing power like a wildfire. Lessig notes, "From borrowers to lenders to Wall Street to government officials—was perfectly rational, for each of them considered separately. It was irrational only for the system as a whole. We need to understand the source of that irrationality—not an individual, but a systematic irrationality" (p. 68). What was the role of technology here? According to Lessig, "with the digital revolution distributing computing power to the masses, masses of financial analysts on Wall Street were able to use this computing power to concoct ever-more complicated financial innovations" (pp. 71–72). Clearly, once the technique was developed and appeared to work perfectly under a controlled environment with a few selected banks, it spread quickly across financial sectors. Individual rationality has now transferred to the collectivity exemplified in a new *cultural rationality*.

Technology and Culture

The attraction toward a *new* technology is not so much due to its availability and accessibility, but to its usage as it permeates our routine actions. Such a realization calls attention to cultural factors beyond intellect and reason in influencing emerging technologies. In introducing the social constructivist theory, Donald MacKenzie and Judy Wajcman (1999) note that the earliest writers who brought attention to the theory of sociology of technology were coauthors William Ogburn and Dorothy Thomas in a 1922 article in the *Political Science Quarterly*. They argued that cultural demands play a more important role than individual cognition in shaping technological progress. Ogburn and Thomas noted:

> Given the railroads and the electric motors, is not the electric railroad inevitable? At least six different men, Davidson, Jacobi, Lilly, Davenport, Page and Hall, claim to have made independently the application of electricity to the railroad. Similar inquiries show that the development of science was leading up to the following inventions, each one of which was invented by several different inventors: the induction coil, the secondary battery, the electrolysis of water, the electrical deposition of metals, the ring armature, the microphone, the self-exciting dynamo, the incandescent light and the telephone. Such a record of electrical inventions, while not negativing [*sic*] the factor of mental ability, certainly shows quite impressively the importance of the cultural factors. (1922, p. 89)

Ogburn and Thomas concluded that culture plays the dominant role in shaping what is socially desired and also dictates that our "ability must be trained and stimulated to attack the problem" (p. 92). Therefore, culture creates the impetus for human ingenuity to be invoked for societal advantage. Culture can be the symbolic incubator that determines what technology ought to be produced and for what purposes. Culture here is used broadly to include the societal values that drive social action.

Culture shapes the human mind and its actions. Without cultural impetus, more importantly cultural awareness, technology fails to create desired social values. Cultures are important carriers of logic and reason that are embedded within our values. Cultural bias toward technical rationality is influenced by the pervasiveness of technology in society (as it directly impacts individual values). Milton Rokeach (1979) in *Understanding Human Values* claims values are embedded in cultures; they are "as much sociological as psychological concepts" (p. 50). Accordingly, Rokeach notes, "Values may be conceived as cognitive representations of underlying needs—whether social or antisocial, selfish or altruistic—after they have been transformed to also take into account institutional goals and demands. In this way, all of a person's values, unlike all of a person's needs, are capable of being openly admitted, advocated, exhorted, and defended, to oneself and to others, in a socially sanctioned language" (Rokeach, 1979, p. 48).

For Rokeach, human values that serve our biological needs (human instincts) are filtered through societal restraints to provide a set of standards to guide us to how to satisfy our needs and thereby "enhance self-esteem, that is, to make it possible to regard ourselves and to be regarded by others as having satisfied societally and institutionally originating definitions of morality and competence" (1979, p. 48). Thus values are appreciative systems constituted and nurtured within a given culture. This can also be referred to as the institutionalized cultural system. They are not static or ideographic; rather, they are dynamic systems of social action because of their "interconnectedness" and "informational" attributes that serve as "carriers of psychological energy" (p. 21). Once the interconnected sets of values are preferred as an "obligatory" state of a social system, they become an ideology (p. 21). Thus cultural values are a *resource* or "psychological investment" (p. 21) that can become handy if applied strategically.

Ann Swidler, in her seminal work "Culture in Action" (1986), argues in a similar vein with Rokeach that culture is a resource, a *tool kit*, for individual action in a society. However, she argues it "is more like a style or a set of skills and habits than a set of preferences or wants" that people can acquire by being part of a society (p. 275). Her emphasis is more on action and practice than on values. Swidler argues that our actions are not guided by our values; rather "action and values are organized to take advantage of cultural competence" (p. 275). In other words, cultural patterns provide the structure that individuals use to develop particular strategies to achieve a certain goal. Strategies are a general way of organizing our action based on available cultural tradition. People living within a cultural tradition have their own interpretation of culture and thus use it as a resource to adapt to circumstances. What endures over time is not one's interest driven by a set of values, but the style or line of action by which one attains the end. In contrast to Max Weber's notion that culture plays an important role in influencing outcomes (his concept of protestant ethic, for example), Swidler notes, "culture provides the materials from which individuals and groups construct strategies for action" (Swindler, 1986, p. 280). Culture is a resource that shapes action, not determines its end. In new nation democracies, ideologies play the dominant role in controlling action, and competing cultural views create new modes of action. In developed democracies, culture is a melting pot with varying cultures encapsulated into one standard guide to behavior. One dominant culture has weaker control over action. In the absence of a dominant ideology, when information technology becomes part and parcel of individual routine, it becomes natural for groups to use technological skills to associate themselves to the culture. The skills and habits associated with technology are routinized as if they are common sense and habitual.

Once the technological habits are part of the culture, they can permeate other habits depending on the value of individualism in society. Bellah (1985) in *Habits of the Heart* argues that individualism is the culture of the Western world. The concept of individualism grows out of a tradition of organizing human behavior—preferences, desires, means, and ends—around an individual. Rationalism has routinized the behavior of individuals as if it is common sense to think rationally and to apply rational systems in all social actions. Rationalism is deeply coded into the cultural system to the extent that it has become a *norm* (legitimized by the plurality) and anything different will be considered abnormal. As Swidler correctly notes, "Settled cultures constrain action over time because of the high costs of cultural retooling to adapt to new patterns of action" (Swidler, 1986, p. 284). Therefore, cultures with a high degree of technological influence should further the individualistic traits within the culture. In fact, evidence suggests that the Internet reinforces the individualism traits through what is now called "networked individualism" (Wellman et al., 2003).

Given technology is ubiquitous, the question of how technology becomes part of the routine culture or value system may appear moot. In the 21st century, we proclaim, "well, it is part of life!" But what role does it play in influencing the settled culture? Indeed, combined with technology the rational culture (the resource) can be used to develop techniques. I argue that technique itself is a cultural tool kit in the information era. If techniques meet the societal value requirements by becoming part of the routine, then they can also become the generator of *predictive values*, a subject that is yet to be explored in the philosophy of technology.

Technology the "Hero"

A cultural system is objectified in images, symbols, language, rhetoric, and artifacts that become part of a group's routine action. The objectification of values is an important part of interpreting and understanding the culture. How I am viewed in society will determine what part I will play in someone's daily routine relative to someone else's role. We are being prioritized by others, depending on where we fit in their routine action. Similarly, technology and artifacts are also assigned priority according to how society values them and how they fit into our routine. We are followers of the cultural system, yet we are different depending on our networks of association based on our routine.

Just as a cultural system can put greater value on a class of people or type of good, it can also devalue people and goods depending on how society *objectifies* them. It is not a rational action, rather a nonrational behavior influenced by a cultural value system. When society puts higher value on objects, others follow the popular culture. We generally objectify computer technology as though it has some mysterious power of solving problems, societal or physical. Jeffrey Alexander (1992) noted that technology has been seen as a "sacred object" that "must be sharply separated from contact with the routine world" (p. 308). Recounting thousands of articles about computers from 1944 to 1984, Alexander noted computers' "charismatic power is repeated constantly in the popular literature," including describing them with words like "Godly," "real hero," "infallible in memory," "incredibly swift," "impartial in judgment," and the "closest thing to God to come along" (p. 309). Such charismatic and utopian views of computer technology have proliferated to the extent that the views themselves are ascribed powers and significance that can influence daily routine and, more importantly, the socioeconomic and political lingo to sway decision outcomes.

The perception that information and communication technologies are charismatic continues to influence the popular sentiment today. Think about owning an iPad, iPhone, or Android (3G or 4G capability), and how people perceive themselves as "connected," "better informed," and "empowered." Just as technology ownership can symbolize status and self-esteem, so do the job titles of individuals managing the information systems: chief information officer (CIO), chief technology officer (CTO) and chief systems officer (CSO). These titles indicate authority, the priority status of those who manage no-nonsense businesses.

Adjectives like "revolutionary" and "transformative" are often used to indicate an intrinsic power within information and communication technologies. The assumption reflects cultural bias toward technology's inherent predictive power for societal improvement. Concurrently, there is the symbolic dystopian view that focuses on the rise of servitude, unhealthy materialism, a decline in family and social values, and the end of human relational attributes (like trust and sharing) due to technological individualism. The discourse is the fundamental reality of living in an open society. Technology is an element in "the culture and the personality systems," and is both "meaningful and motivated" (Alexander, p. 314). What is more important in the trajectory of the discourse? We need to judge information technology in its own right as a resource for the purpose of human and social development. With technology's liberating force comes different costs, so we must be mindful of minimizing those costs to society. For this, we need to gain a deeper understanding of technology's integration into routines and its association with the social network.

Performance, Action, and Actor Networks

The discussion so far has highlighted an important phenomenon with respect to technology: It is considered a critical tool within the culture; it is part of our association to help us interpret the world differently. Hence it has been given a status just as any other authority. Therefore, it changes the ontological status of the broader role of technology in society.

I come back now to the initial discussion of society as a social construct that we are able to define and understand only from a relative stance given our interpretation of it (Ihde, 1990). It is critical that we take this approach because it helps us see how technology affects society and how society is affected by technology—how they are co-constituted. The relative understanding of society shifts the discussion toward a paradigm where society can be perceived only in its own terms, not in terms of cultures or how others may define it. In other words, we can only permeate society by connecting the dots of association—one actor to the other, be it human or technology—as it is exemplified in action. Rather than actors entering into a society already defined with its given rules and structure, society is constructed by connecting the *relational* pieces together to form a network of association as found directly from the attributes of the *actors' actions*. Hence, as scientists we can only study what holds society together by collecting documentary evidence indicating how one entity relates to the other. We must observe, listen, and patiently watch how connections evolve into an active network of association as defined by the actors' own rules of association. According to sociologist Harold Garfinkel (1967) this hands-off approach allows us to avoid self-constructed (imposed rationality) cultural labels and begin to perceive the transformations of social actors by investigating how they occur. He was more interested in studying the commonsense and localized situational context than in perpetuating the outsider's perception of local behavior. There are social worlds within the social world, each with its own embedded rules and regulations that can be learned by studying the association of actors within each social group. Garfinkel's work illuminates the concept that technology is part of the social world and is enacted within the social realm by individuals.[6] The applications of his ideas have been extended to argue a domain independent ontology of association known as actor network theory or science, technology, and society (STS).

The idea of actor network pioneered by the work of Bruno Latour (1986, 2005), Michael Callon (1986), and John Law and Callon (1992) argues that society is not a collectivity of homogenous groups of people and objects that associate and connect to other smaller or larger groups (macro/micro distinction). Rather, it is a heterogeneous, yet seamless, continuous, and fluid assembly flowing over a landscape. An assembly can be identified as a group when it appears to *stabilize* due to its performative action over a common resource pool, until the assembly disintegrates to create another set of associations. Those associations may have already been present or developing until we identified them as another group because of some common performative action that caught our attention. Latour explains: "The world is not a solid continent of facts sprinkled by a few lakes of uncertainties, but a vast ocean of uncertainties speckled by a few islands of calibrated and stabilized forms. . . . We have to be able to consider both the formidable inertia of social structures and the incredible fluidity that maintains their existence: the latter is the real milieu that allows the former to circulate" (Latour, 2005, p. 245).

What Latour essentially argues is the formal world is floating within the informal core. It is the formal world that we can objectively identify and measure, yet it is the informal traces that help us find the meaning of the formal world as it evolves intermittently. Essentially, Heidegger's conceptualization of "revealing the truth" from the "standing reserve" is a close read as described by Latour as the network of association. Latour argues that there is no formality involved in the process of revealing the network of association, since when actors enter and leave networks of association, they *inscribe* to new objects, artifacts, or even plans, then *translate* that to open new avenues for association. The inscription and translation is the temporal state that is corollary to Heidegger's (1977) idea of *enframing*, which is a "state of being" that encompasses a time and place that destines someone to discover or reveal what was yet unknown.

Latour alludes to an operational concept that guides us to see the transitions beyond time and space. *Translation* is to form the idea, and *inscribe* is to formalize the information into an artifact such as information technology. What is important in the idea of inscription is that inscriptions "increase either the mobility or the immutability of the traces" (Latour, 1986, p. 11). In other words, inscribed objects can become mobile (transferable from one person or location to the other) while preserving the originality of the context (time, place, and history). Therefore, inscriptions such as information technology can be both constraining (immutable) and enabling (mobilizing). Constraining in the sense that technology reproduces the "original" without being able to recontextualize with the emerging situations. It is meant to reproduce the same as a routine action (by design) so that, for example, the same database structure that was created to collect and analyze data for one agency may be used for multiple agencies across the globe. Per the objective of the technology, what is mutable is also mobilized through it. Even if the database is inappropriate for the agency, its inherent, automated, rational behavior to reproduce is enough to attract managers for the sake of being efficient and organized. It becomes the standard, or "perceived truth," because of the immutability and mobility of the replicas. In Latour's words, mechanisms have been invented that, no matter how "inaccurate the traces might be at first, they will become accurate just as a consequence of more mobilization and more immutability" (Latour, 1986, p. 12). Therefore, when we find an activity useful for us to "carry-on" with our informal routine, we try to capture (inscribe) the essence of that activity so we do not have to recreate it (immutable). Once it is inscribed (written on your outlook calendar or notepad for example) we find ways to mobilize it again and again so that it becomes part of our daily routine to help us "carry on" with our lives.

Inscriptions are powerful interpretive tools because they are objects one can visualize and they feel more than just hollow imagination. Because of that objective proof, they can be "sold" to others with clear objectivity and decisiveness through texts and graphs and statistical techniques. However, if the formative inscription cannot be universalized, it becomes immobile and cannot be translated to other groups or cultures because it could not be standardized or universalized, and the scale of domination through inscription diminishes. Therefore, for some artifact to be appealing, it must have the dual feature of immutability (inscribed in the principle) and at the same time transferability (mobility through standardization and automation). One without the other will make the technology less appealing and eventually obsolete.

What is immutable is the quintessential element of technology. It answers the core question: Why am I going to use this? What purpose will it serve for my organization? Going back to the previous example, given the fact that the database was created for a particular agency in mind, it may work well for that agency and for those agencies that are exactly the same. However, agencies are not run by machines; they are not exactly the same. Human-technology relationships will always produce different results because of the human variable that cannot be predicted or controlled. So when the technology is transferred to another agency in another context, its effectiveness will never be the same. It could be better or worse depending on human creativity, opportunity, organizational context, and knowledge. In other words, we are able to take the information but leave the traces behind. Using the same information, new traces are produced. The more creative (more customized) ways will make it less mobile for application in other agencies. The tension then is to tease out the immutable part of technology enactment (assuming the whole process as a structure or artifact) that should be mobile and potentially useful regardless of context. How do we move forward? What part of immutability do we leave behind because the parts are old, applicable to different times, different people, and different contexts?

Resource Mobilization

So far we have discussed inscription and human action. Technologies in general are inscribed artifacts designed to reproduce repetitive and routine action. We have an indirect understanding of what technology is or what it could be from the cultural understanding of its power and usage, discussed earlier as the perceived understanding of technology. There is also another understanding of what it is based on: our direct experience when using it. Both of these understandings reflect on how technology is enacted or performed. Through performance, the abstract world becomes concrete. This *performative* improvisation depends on individual cognition, imagination, and creativity. Once the performance of a technology is enacted within individual routine, it becomes part of the performative action within our social world.[7] The performative action differs not only due to an individual's cognitive ability and cultural system, but also due to the situation or context in which it is being applied. What Garfinkel calls the "normal, natural troubles" (1967, p. 192) are part of reality; no matter how much control we may try to impose on given activities they will produce routine yet imperfect outcomes.

Brian Pentland and Martha Feldman (2008) use narrative networks to argue that performative actions occur within live routines as opposed to dead (immutable) routines. Whereas dead routines are "rigid" and "mindless" artifacts, live routines involve "people who are capable of learning from experience" and are "generative," giving rise to "new patterns of action" (p. 240). Managers who rely on rigid routines, either consciously or subconsciously relinquishing management to the design of the hardware, can kill the routine through total automation. The discretionary authority of users significantly decreases when routines are equated with techniques, a fixed way of doing things. The techniques are often subtle, appearing in formulas, flow charts, forms, and rules. These techniques or dead routines also appear overtly in workshop PowerPoint presentations and mandatory reporting requirements (possibly with legally binding enforcements) expecting the same results again and again like a chemical reaction of two chemical components. It is easy to see how dead routines would be preferred to live routines in organizations that "impose" outcomes by rational and standardized actions. Routines are stifled when artifacts incapacitate humans; however, routines are generative within environments that "allow patterns of actions to be produced based on local judgment and improvisation by actors" (Pentland & Feldman, 2008, p. 249). Any technique and technological routine will certainly generate an outcome, but it will always be at the expense of the people who will be affected by it unless the techniques allow customization to local "live" routines.

In an organizational context, the practice that comes out of the live routine redefines the relevant *resources* available to the organization that then *mobilizes* the social system. Feldman (2004) and Orlikowski (2000) emphasize that rules and resources within an organization become the essential part of the narrative. Once resource use is routinely practiced in a certain way in relation to a particular context, it stabilizes, and the process reduces into a schema—"technology in practice" (Orlikowski, 2000, p. 409; Feldman, 2004, p. 296). The actions themselves create resources that can be used, reused, and mobilized. The continuous enactment of resources through our actions produces schema. Wanda Orlikowski (2000) gives an example of a schema that is enacted during the tax filing season just before the April 15 deadline, "when users are motivated by the tax filing deadline to use their tax preparation software in a flurry of repeated activity and anxiety, and thereby enact a particular technology in practice" (p. 408-409). The concept of schema is applied to human situational context with respect to the technology in use. Whereas a given technology is not under control as far as its physical attributes (automated action by design) are concerned, the human situational context, the schema, depends largely on the resources that enhance or impede the human situational attributes. Put simply, resources available to actors affect their abilities to enact schemas. Therefore, technology becomes the enabler for activating the resource-schema continuum that affects the outcome of any interaction between human and technology. Resources here are invaluable because they are the "core of enactment" that can be mobilized for others' practice with technology.

Two types of resources are fundamental to enacting the schema: the relational resource and the informational resource. According to Feldman (2004), the relational resource includes "relational qualities such as authority, trust, connection, and complementarity of people" (p. 300). The informational resources are "expertise and organizational knowledge," that, according to Feldman are "assets that are implicated in the resourcing picture" (p. 300). Whereas relational resources grow informally within our routine, informa-

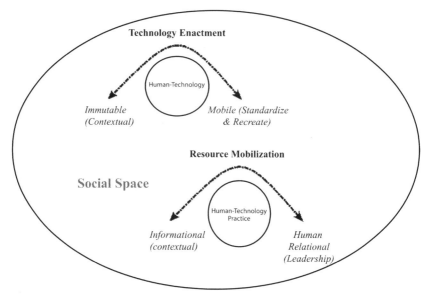

Figure 4. Human relations technology enactment

tional resources are gathered through formal protocol and legitimized by designated authority. Feldman has shown how an organization's relational resource, coupled with the informational resource, can be critical to how organizational schemes are enacted and how this combination affects action. Since action drives the resources, the relationship between resources, schema, and action is recursive and nonbinding; one does not follow the other. Because of the recursive nature, the process is nonlinear at best.

Human Relations Technology Enactment

The above discussion can now be extended to a theory of human relations technology enactment (see figure 4): Technology as an artifact is both immutable and mobile. Immutability is the core principle of technology enactment. It is contextual in nature and generates performative (live) routine action. Part of the routine action is a knowledge resource by itself that helps us understand how the particular technology is enacted under what circumstances. That knowledge resource is mobile and transferable. This mobile knowledge resource should not be confused with technology transfer, transferring a machine from one office to another office, or creating a database and selling it to another organization. Rather, it is the practical knowledge of technology enactment that is argued here as the resource for mobilization related to the particular technology and context where it is applied. Information relation and human relation combine to embody the transferable resource. Information relation is directly tied to the knowledge and technical expertise of the technology that will be applied in the new context. I argue that the information must go beyond technical expertise to an understanding of the context in which it will be applied. Context itself is the interpretive tool that should be the platform to assimilate information through technology. Therefore, it is directly connected to the human relations resources like trust, networking, democratic values, ethics, responsibility, and accountability that reflect our desire to build and use technology for human advantage.

Conclusion

The implication of the ideas discussed so far highlights one important fact: technology is constitutive; its formative nature can be valuable when it is left open to human imagination and improvisation. The conceptualization that technology is constitutive socially and culturally is of significance for building stronger democracy. We have focused here on the resources that enact our technology practice because relational resources are fundamental to our understanding of public technology and our role as public administrators in a democratic society. Some scholars in the field have idealized technology as merely a neutral tool focused on the relational resources from the point of view of building social capital. Others capitalize on the informational resource with the importance of building the knowledge economy. The relational resource and the informational resource are not mutually exclusive; they are constitutive and reflected in the embodiment of technology enactment. In a technologically dominated society we have focused on technology and not the resources that enact it. In a digital democracy, we have so far failed to see the light of technology through the better lens of humankind. We must develop and nurture the resources that enact technologies' formal nature to the informal human routine. The challenge is hereby presented to us, and we must prepare for it with full force.

Part II
Value of Public Service

3

Information Contextualization

Imagination is more important than knowledge. Knowledge is limited. Imagination encircles the world.

—Albert Einstein

Whether information is formal or informal, quantifiable or nonquantifiable, we gather it sometimes consciously, but more often subconsciously, in order to *move on*[1] with our natural social action. Our natural social action helps us maintain and build relationships based on our values. There is no rational calculative motive in our social action; it is spontaneous, natural, habitual, informal, and mundane. We are part of the *flow* guided by our values to continue our social relations that we have built trust upon, and we hope to extend that relationship to larger social and natural (earth and living) groups. In the process of building relationships, we have to be informed as to how we can continue to maintain those relationships based on what we already know and trust; at the same time, we must be informed as to how to build new relationships that conform to our values, and then we must adapt to those new relationships. Therefore, in order to move on with our natural social action, we have two tasks at hand: (1) How to maintain what we already know, and (2) how to travel into the realm of the unknown. For the former task, we employ formative tools to formalize the structure so that we can take control of the already known information and maintain existing social relationships; for the latter, we need to learn—through vigilant eyes, imaginative tools, and an open mind—so we adapt to emerging values and expand our social relationships. For the former, we may settle with habits and techniques for maintaining control of our existing social relations; however, for the latter, we have to learn from the emerging situations by reconnecting with the past from what we already know (historicity). Here we must "imaginize" (Morgan, 1997) and contextualize new (unknown) information to understand emerging values. *Dry* and noncontextual information may be sufficient to maintain the existing relationships. However, in order to understand the values that are outside our known social relations, we need *wet* information that gives us insights from the context and practices of individuals outside our social relations. As Lave and Wenger (1991) noted, learning the ropes of a new community requires "legitimate peripheral participation." We learn through the gradual steps of being made a fully participating member of a community, not through information stripped of context.

In the last two chapters, we discussed how information is formalized and inscribed to create routine action. Because formal information is immutable and inscribed into some physical structure, it is transferrable. The characteristics of formal information are to be immutable and mobile to allow information to reach out to situations beyond its origin. Informal information that is not capable of being inscribed due to its nonquantifiable and value-laden nature escapes most inscriptions unless we are conscious of them. In other words, informal information that is continuously emerging due to changing times, situations, and context has no way of being fully captured unless we have a filter that consciously scans the information to learn and add value to what we already know. This chapter deals with informal information—how it can be contextualized with formal information and what role information technology can play in contextualizing information so as to establish new social relations. As argued here, information technologies can be used as interpretive tools to construct new social relations.

Structuration Theory and Contextualization

No information is useful without imagining how to strategically use it for managing organizations (Morgan, 1997). To *imaginize* without context is to dream it without the knowledge of how to bring it into reality. In order to explicate the context, we must understand the duality of structure (Giddens, 1984; 1979), which is the basis of how we take given information that is already tested (forming the "structure") and *contextualize* it with new information by observing others in practice. Information technology that has become the predominant source for all our information can be a practical interpretive tool to use to discover the value of the new information. Context is often understood to (hypothetically) ascribe the meaning of one situation with another, based on differing circumstances (time, place, and history). Another operational meaning of context is that it is a discourse where unknown information is interpreted based on known formulations. In other words, the "inner experience" of the actor is *sensed* using "outer world" knowledge. We increasingly have a *better sense* of the inner world when we expand on our outer world experience. As the outer world expands, our distance to the inner world decreases. This phenomenon can also be explained in terms of rule ordering. When we understand something to be the norm from our inner

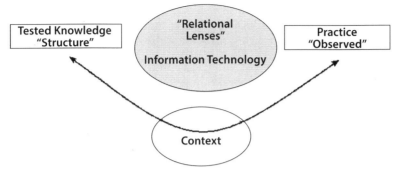

Figure 5. Building information from context using IT

experience, it becomes part of a "formal" rule ordering. We judge the outer world based on that rule ordering (experience). In fact, we do not have anything more standard to evaluate than our own experience. The outer world experience can be viewed as an object "constituted through some 'unifying' principle, some logical operator into an *order* of relationships to each other" (Garfinkel, 2008, p. 133). Establishing this *relationship* of the unknown information (outer world) to that of the known information (inner world) creates the context. The relationship is established by using selective criteria that match the already known formulations while neutralizing those that do not appear to have any corollary. The relationship grows as we expand our selections to include a wider pool of information. This could be achieved directly through firsthand experience of getting involved into diverse areas of professional and nonprofessional activity, and also by expanding our primary social group. However, given the Internet's presence in our daily lives, the informal information of the outer world is now within our reach. Information technology can become a new *interpretive aid* that can enable us to contextualize known formulations with that of the unknown. What is profound in the subject of contextual information is it provides insights to the values of others that are foreign to us and allows the given values to be constituted with "socially validated realities." This expands the "communicative net" for improving mutual understanding (see Ann Rawls' notes, Garfinkel 2008, pp. 72–83). Indeed, democracy is strengthened when we exercise a diversity of values. Figure 5 explains how context mediates between the inner world of tested knowledge and that of the outer world's observed practice. In this mediation scheme, information technology is a socially active conduit that provides the tools (relational lenses) to use to interpret emerging values and their relationship to the existing values of a society.

The theoretical underpinnings of contextual mediation can be explained using Anthony Giddens' structuration theory (1979; 1984). The theory of structuration recognizes humans as knowledgeable agents who actively engage resources, both formal social norms and tangible artifacts, to make things happen. Structuration is a dynamic process embedded in historicity, both time and place, giving rise to formal rules, and to how a particular act is performed (that is, practice). This *duality of structure*, as coined by Giddens (1979), in essence can be described as the reconciliation between the *syntactic* world of formal codes and regulations (objective), and the *semantic* world of practice (subjective). Our actions exemplify a duality that can be best explained by formal language and our (informal) usage of that language. The informality of the language can only be realized in practice when we interact with others and learn how others use it in daily life. In this sense, context is central to how this duality will create new meanings. As Giddens (1979) noted, "*The context of interaction is in some degree shaped and organised as an integral part of that interaction as a communicative encounter*" (p. 83; emphasis in the original). The context is the *platform* where "reflexive monitoring of conduct" takes place. The reflexive monitoring is the *accountability* of what one *does* based on what one is *supposed to do*. What is reflected in practice is therefore the accountability of the formal norms, reflected in our actions and inactions. Structuration theory argues that social norms are factual boundaries of social life. Formal tested knowledge acts as a constraint to our everyday action. Therefore, on the one hand, our actions are *constrained* partly by the historical binding of formal norms and rules that signify how we ought to behave in our social interactions (setting the moral order); on the other hand, our resources *enable* us to continue with our social relations in ways that exemplify how goals will be accomplished and how power will be exercised in practice (setting "observed practice"). The outcomes of these interactions are neither predetermined nor predictable because they vary depending on the power of the *interpretive schemes*. Interpretive schemes are "the stocks of knowledge which actors draw upon in the production and reproduction of interaction," they are considered "the same as those whereby they [the actors] are able to make accounts, offer reasons, etc." (Giddens, 1984, p. 29). Information technology, in recent years, has become one of the most powerful aids used to materialize human imagination by processing together the information contained within a human mind and what is gathered from societal interaction. Internet-based information technology today offers us a clear choice between expanding our *interpretive power* through contextualizing to *inform social relations* or continuing to use it as an "instrument of domination and control" (Law, 1991). Information technology as a "medium of human action" (Orlikowski & Robey, 1991, p. 161) can expand our stock of knowledge to expand our imaginative and creative power. Particularly, social media tools, the blogosphere, open multimedia platforms, You-Tube, Vimeo, and the Internet in general should no longer be viewed as instruments or as means to achieve desired social goals. Rather, modern-day information technologies' far-reaching *social relational* attributes must be acknowledged. These attributes have the power to humanize information by resisting the crystallization of forces into the asymmetries of power. This may sound radical, yet the resource is before us everywhere, what we say and how we say it, is how we are informed. Indeed, "Technology is a mode of revealing. Technology comes to presence in the realm where revealing and unconcealment take place, where *alētheia*, truth, happens" (Heidegger, 1977, p. 13). Therefore, technology can be viewed as a relational tool that helps us *see* and even feel (or empathize with) what may be discovered in others' practices.

As social relational tools, ICT can be viewed as *mediators* as opposed to *intermediaries*. Unlike intermediaries, *mediators* transform the message as opposed to just transporting it without distortion or addition.[2] With respect to mediators, when a message is being transported, it is customized based on local context. Therefore, mediators impede standardization of the message as it finds ways to channel the "moral" message through improvisation to address the needs of the day. Intermediaries are the primary vehicles for creating "black boxes" or closed systems where the input leads to a given output and the interlocking of the coordination between input and output is not clearly identified (Kaghan & Bowker, 2001). Clearly the Internet and ICT in general are effective intermediaries, but they can also be powerful mediators when used to disseminate situated information for social transformation (Haque and Matonde 2013). Indeed, the view of ICT as an interpretive tool can only facilitate the development of new technologies that will advance the technologies' contextual capabilities, thereby improving the relational lenses used to expand as well as liberate our interpretive power.

With all of this in mind, actual information can frame our mind only through some experience. In other words, information is constitutive from practice. What is known as "gut feelings" that are used to make a decision can only be gathered directly from locals on the ground unless they can be somehow transferred and translated to describe the actual practice. This is where ICT is becoming far more valuable. ICT is becoming increasingly important as representational technologies or interpretive tools that have traditionally been underutilized and even ignored at the expense of expanding our stock of knowledge for better decision making.

Information Contextualization Tool Kit

The Narrative Inquiry

Narratives are stories that are thematically related to a sequence of events, experiences, or actions to create a meaningful whole about the information presented by an individual or group. The storyteller's ideas must connect many unknowns and somehow convey their significance and meaning. The digital narrative world is expanding. For example, Tumblr.com, a social blogpost site started in 2007, has more than 101 million blogs with more than 44 billion blogposts. Why do people want to share stories? A narrative is an expression of an individual's expectations of a hitherto uncertain and unknown world that can translate the expression into meaningful information. It is a constructive communicative tool used to make sense of what is being created out of sequential events that are linked to other people, space, and time. Narratives are useful to the narrators and to the observers because stories carry information relevant to decision making (Feldman et al., 2004). They identify network and power relations (Nespor & Barylske, 1991) and strengthen deliberative democracy (Polletta & Lee, 2006). Narrative inquiry is an important route to knowledge for practitioners who can reflect situated social reality from people's intentions, beliefs, and emotions. The narrative draws people in a "series of complicating actions" that, once understood relative to their own familiar stories of "loss and enlightenment," are now assembled in a way that enables them to identify "imaginatively in experiences quite unlike their own" (Polletta & Lee, 2006, p. 713). This narrative inquiry can be called reflexive practice, where we open our senses to outer experience for learning by example.

For public administrators, narrative can be one of the most powerful communication tools to create shared meanings and shared understanding that brings public discourse to public management (Miller, 2012; Hummel, 1991; Ospina & Dodge, 2005a; Stivers, 2002, 2011). Although the root of narrative inquiry in public administration can go back to the work of Philip Selznick (1957) and Chester Barnard (1968), it has received important currency in the mainstream research of public administration in recent years (Miller & Jaja, 2005; Dodge et al., 2005; Ospina & Dodge, 2005a, 2005b). One of the important implications of using narrative inquiry is its communicative power. It is a communicative vehicle that practitioners can use to establish "solidarity and trust" in government (Polletta & Lee, 2006). It also enables practitioners to build leadership qualities for social change within marginalized populations whose world generally remains unreached through traditional methods of inquiry (Ospina & Foldy, 2009). It is a cultural tool kit that individuals can employ to answer the often-asked question: "What do you mean?" Narratives in fact have been underutilized and underappreciated because they appear less scientific (Hummel, 1990; 1991), and they cannot be reduced to mere standard codes that can be transported and mobilized for scientific investigation. Regarding the usefulness of narratives, Hummel (1990; 1991) explained that an "object to be investigated must first be constituted before it is scientifically analyzed" (Hummel, 1990, p. 303) and that "reality is constituted not by consensus of all imaginable detached observers but by the present community of those involved in a problem who must be brought along to constitute a solution" (Hummel, 1991, p. 33). Therefore, narrative in essence is the traces of information—as a vessel it holds the quintessential message that must be carried forward to write yet another story, for another time and place.

The credibility and value of narrative inquiry has been questioned due to our own cultural biases of what we think society ought to be. The cultural biases that go with narratives can create doubts about the credibility of a narration or deliberation. People's socially endowed capacities are different (see Bourdieu, 1990), and we expect their deliberation and narration to be different. Yet people are reluctant to expose the differences and go against the habits and norms that are dominant symbols of social status. We self-perpetuate established habits by conditioning ourselves to group norms and remain *suppressed* in our own social bounds. We deliberate in public in a way that conforms and reinforces the accumulation of "symbolic capital" so that we sound credible within our social group, limiting our ability to assimilate or become accepted in other social associations. Thus our personal narrative is overshadowed by the sheer dominance of the self-regulating *habitus* of social order mediated by objectified or even institutionalized mechanisms. In Bourdieu's words: "Social formations in which relations of domination are made, unmade, and remade in and through personal interactions contrast with those in which such relations are mediated by objective, institutionalized mechanisms such as the 'self-regulating market,' the educational system or the legal apparatus, where they have the permanence and opacity of things and lie beyond the reach of individual consciousness and power" (Bourdieu, 1990, p. 130). If technology is institutionalized to create routine habits, it works as an intermediary (input-to-output) only to reinforce the control mechanism. On the contrary, if technology plays the role of a mediator, it allows human intervention, from simple improvisation to deliberate reflexive action, depending on one's practical knowledge.

In the online world, digital narratives in the form of digital diaries (blogs, wiki, and social media) and even e-books have opened up unprecedented opportunities for personal narratives that challenge the universality of ostensibly universal principles and push the envelope for a new deliberative agenda. Digital narratives are indeed artifacts; they are immutable yet mobile constructions, each with a life and longevity of its own. These textual artifacts are seen as communicative manifestations of culture; they endure and give historical insight (Clair, 2003). Modern-day artifacts have become important tools for the self-expression and mobilization of social groups who had been marginalized from traditional public discourse due to many obstacles including social class domination. For example, graffiti narrates the expression of phenomenon and removes the physical boundaries of domination (Scheibel, 2003). Increasingly, the online blogosphere is becoming a platform for self-expression and imagination of the world around us. Stephen Coleman (2005) argues blogging should be taken seriously because it is becoming a "democratic listening post" for politicians and bureaucrats. Coleman notes, "Blog is to declare your presence; to disclose to the world that you exist and what it is like to be you; to affirm that your thoughts are at least as worth as hearing anyone else's; to emerge from the spectating audience as a player and maker of meanings" (p. 274). Even for marginalized communities, it builds the capacity to govern and becomes "a source of nourishment for a kind of democracy in which everyone's account counts" (p. 274). Despite the digital divide, blogging is not limited to small groups—the blogsters are as diverse as the globe itself.

Narrative is neither intrinsic to any technology, nor is it a modern-day fad. It could be any artifact that helps express one's hopes and imagination through language. According to Gadamer (1975), there is no such thing as objective knowledge; all our understandings are embedded into situations that encompass time, history, and tradition. He argues that the medium of language is a powerful tool with which to interpret reality, which explains phenomena with historic authenticity. Citing the work of Hans-Georg Gadamer (1900–2002), Giddens (1993) argues language is "the ontological process of human discourse in operation" and that it "does not comprise a procedure on interpretation, to understand language is to live in it" (p. 63). Essentially, language is thought of as an unfolding of experience. For example, throughout centuries, though little known in the Western world, the embroidered quilts in Bangladesh and West Bengal, India, have been the expression of women's empowerment. These very sophisticated *embroidered murals* on long sheets of cloth (*sarees*) known as *Nakshi Kantha* have been a platform for women to use to express their feelings and dreams through weaving-narrative. Rural women who could not otherwise read or write found subtle ways to employ their own skills to express their imagination and fears (see Ahmed, 1999).

Bengali poet Jasīmauddīn (1903–1976) wrote about *Nakshi Kantha* (embroidered quilt) using metaphors of a young girl named Rupa:

> Spreading the embroidered quilt
> She works the livelong night,
> As if the quilt her poet were
> Of her bereaved plight.
> Many a joy and many a sorrow,
> Is written in breast;
> The story of Rupa's life is there,
> Line by line expressed. (Jasīmauddīn & Milford, 1939/1958)

Bruno Latour (1986) notes that proper perspective is required if one is to make the translation of reality undistorted. Interactive maps with live YouTube discussion allow us to visualize, hear, and observe the relationships that are established in the new and unknown association: "The speakers are talking to one another, feeling, hearing and touching each other, but they are now talking with many absent things presented all at once. This presence/absence is possible through the two-way connection established by these many contrivances—perspective projection, map, log book, etc.—that allow translation without corruption" (Latour, 1986, p. 8).

Scholars in public administration have argued that films, arts, and painting are powerful narratives that provide deep insights into *human relations* (McSwite, 2002; Marshall, 2012; Chandler & Adams, 1997; Goodsell & Murray, 1995). Films connect individual processes to larger social processes by translating our own images of self into an interpretive social reality. Our response to what a director creates in a melodrama is more about our own theories involving the society being tested. The melodrama will either move us, because we draw inferences from it given our theory, or it may just entertain us as a passing play. We can relive the experience even if we have not traveled on the same path before. In this regard, films and even paintings can be powerful media for connecting theory to practice.

Photovoice Method

A subcategory of narratives known as photovoice—narration by photographs—has recently become an important community-based research tool within the health care practitioner community in the United States. Initially developed by Wang and Burris (1994) to enable Chinese village women to photograph their everyday health and work realities, the photovoice concept allows participants to "control the photographic process in order to express, reflect and communicate their everyday life" (Wang, 1999, p. 186). It is a community-based action research strategy that allows understanding of community assets and needs for empowerment through photographs and pictures. Photovoice was particularly geared toward feminist inquiry to eradicate the disparity between how women are seen and what they feel from their own portrayal of life as they experience it every day. Each participant in the photovoice project is given a camera to photograph or permit another to photograph events of their daily life. The pictures can be anything that the subjects want to portray as representative of their daily lives. Quoting psychologist Abigail Stewart, Wang (1999) asserts photovoice is about studying women's lives by "what ha[s] been left out," the concerns that have been "overlooked, un-conceptualized or ignored yet may be central to women's experience" (Stewart, 1994, quoted in Wang, 1999, p. 190). The photos are then defined and interpreted for a live audience of policy makers and practitioners. According to Wang (1999), health care practitioners appeared to be far removed from reality. She noted, "it was the health policymaker who was often far removed from the experience of the people whom they govern—who could and did learn from women photographs and stories" (p. 187). The organizers of the photovoice project must abide by an ethic of responsibility to give the subjects full authority to take or share any and all photographs. Most importantly, the pictures are owned solely by the participants and are returned to them when the project is completed. The photovoice approach is particularly useful when used to understand phenomena such as life in a homeless shelter, elderly health and health behavior, women's health, and the lifestyles of marginalized populations who do not have access to public participation or whose lives cannot easily be captured or objectively understood using formal research tools. Among health professionals, photovoice is gaining wide appeal as a community assessment tool for understanding the socioeconomic implications of health and resilience issues related to cancer survivors, HIV prevalence, obesity, teen pregnancy, environmental hazard, and risk assessment (see, for example, Alegria, 2009; Catalani & Minkler, 2010; Hennessy et al., 2010; Lardeau et al., 2011; Neill et al., 2011; Tanjasiri et al., 2011). Its usefulness as a practice-based community assessment tool has received national attention from the National Institutes of Health (NIH), including the National Cancer Institute (NCI).

Geographic Information System (GIS) and Visualization

Cartographers use the term "graphicacy" to describe "the ability to communicate effectively and understand those relationships that cannot be expressed solely with text, spoken words, or mathematical notation through the use of visual aids, particularly maps" (Hallisey, 2005, p. 351). Visualization can be a powerful way to understand and communicate about people and the surroundings that influence their actions. It is probably impossible to capture all relevant attributes about people using any known method. However, given the digital visualization of people and places, we can use GIS maps to construct an interpretive framework by which we explain people and related phenomena and how our decisions affect a particular group or community under investigation. Human geographers emphasize, in particular, the use of holistic techniques to understand the social process of human interactivity and behavior.

Although GIS is what geographers have used for decades, its appeal when used to capture the activity of people and phenomena through visually appealing maps has greatly influenced decision making beyond geography to almost all aspects of social, physical science, and human geography today (Lamb & Johnson, 2010; Elwood, 2010, 2011). According to a survey conducted in 2004 by the US Department of Interior, 97% of the large local governments use GIS for critical tasks such as planning and construction, transportation, crime tracking, public access information, law enforcement, and emergency preparedness (McGill, 2005). The US Department of Labor lists GIS as one of the leading skills to meet the demands of high growth industries. By incorporating space or location into decision-making models and displaying the attributes or characteristics of the space, GIS can be used to create different scenarios (or maps) that any layperson can understand. Differing views can now be brought to the table for discussion, and all stakeholders can visualize their stakes by looking at a set of maps with "visually meshed" multimedia links to places on the maps. Target-based use of GIS is recognized as the most effective decision-making tool (Sliwinski, 2002; Thrall, 2002). It is arguably one of the most interdisciplinary and collaborative tools used in higher education today (Kawabata et al., 2010). GIS finds its most use in solving complex problems and finding nondeterministic outcomes. Although GIS is not immune to the rational bias embedded in any technological system, allowing the individual to view a phenomenon "as is" on the ground gives the ability to interpret and discover issues that may not be apparent in a single database model or information system. Once the information about the phenomenon can be captured through satellite, aerial, or vector-based images, it can be used for contextualizing scenarios about how or where a given population might be affected, or who ought to weigh in on a final decision. Not only can ideas be generated much faster but also a large number of people can participate in decision making based on one common theme depicted visually on a GIS platform. These visual representations can provide insight into the similarities and differences among scientific concepts and a priori or even stereotype assumptions held by people within the academic community. Clearly, visualization using GIS can allow groups of researchers to share and refine concepts and even negotiate common concepts. Unlike any other time in modern history, having disparate datasets in one organized system allows managers to have access to varied citizen information on their desktop or handheld devices. GIS is not immune to problems that are inherent to all information technologies. Just as pictures can be worth a thousand words, they can also tell a thousand lies (Monmonier, 1991). They can be misused and manipulated or even used in ways that can undermine democratic values (Haque, 2001, 2003).

Recently, public administration scholars have used GIS to reveal how public administrators have been used as the very means to desegregate and dispossess African American communities (Alkadry & Blessett, 2010). GIS has been found to be the best tool to use to address issues of social and environmental justice (Mennis, 2003; Boer et al., 1997). In addition, GIS is being used to understand changes across the economic, political, and social landscape. The number of Internet sites dedicated to GIS is too exhaustive to list here. One of the examples that can be highlighted is in regard to the migration of population from different counties of the United States in 2008. The webpage "Map: Where Americans Are Moving" (Forbes Magazine, 2011) uses IRS intercounty move data to visually demonstrate migration flow across the United States. GIS websites have also been dedicated to making government more accountable by tracking stimulus spending allocated to each state through the American Reinvestment and Recovery Act (ESRI, 2011a). Web-enabled, user-generated maps such as those provided by ESRI ("Mapping for Everyone") can be a powerful way to engage students about policy mandates and implementation issues (see ESRI, 2011b). The federal government, through its Data.gov: Empowering People (US Government, 2011) initiative has listed links to all federal government departments that have publicly available geo-referenced data for GIS mapping. It is instructive for public administration students to engage in exploring and constructing contextualized case studies of federal agencies using geo-referenced data with available visualization tools.

Google Earth, Bing, and Social Media Mash-Up

In the last few years, we have entered a new era of human-technology interactivity using web-enabled interactive platforms. For instance, a few years ago people used Google Maps, MapQuest, and other online maps to find directions. Today, web-enabled maps such as Google Earth and Bing are used to mash up geo-referenced (location specific) pictures, stories, and live videos attached to locations anywhere on the globe. They are becoming powerful public educational tools for teaching and learning in classrooms (Lamb & Johnson, 2010). Bing maps introduced features such as oblique imagery (Bird's-Eye View), Streaming Video, and Photosynth that allow individuals to have a wide understanding about a community. Bing's newly added "The Education Map" helps locate matching K–12 schools that meet particular requirements as well as mentoring and volunteering opportunities in a community. "Data Market Mapper," another Bing app, connects specific socioeconomic data of interest reflected on the maps. The large amount of geo-referenced contextual data clearly highlights the fact that students today have a wide variety of opportunities to gather field-level data directly from publicly available sources and use it to share and generate ideas and correctly interpret them to find appropriate solutions. Students can pick an area of interest, conduct virtual community assessment, conduct fieldwork on the ground ("ground truthing"), and share stories online through pictures, videos, and texts. Online narratives can empower collaboration and facilitate shared processes to generate ideas to solve problems. Contextual maps allow us to comprehend differences and the diversity of people and places whose varied contexts present varied challenges. According to Terry Cooper (2011), maps and videos are powerful civic engagement tools that can be used for "generating deliberation, building support for certain policies or policy changes, and communicating with administrative agencies" (p. 351). Cooper argues, "Significant progress in democratizing the administrative state through citizen-driven public administration is within our grasp," and that "sustained engagement" will "fulfill the promise of democratic governance for our time" (p. 254). Indeed the 21st-century generation should be ready to take on this challenge and use technology to contextualize what they see and learn.

Science and the Art of Decision Making

Information contextualization as a method of interpretive inquiry does not receive wide credibility compared to traditional "scientific" methods. Traditional scientific inquiry receives wider credibility and acknowledgment due in part to its objective quantification of context. This is partially because of Western bias toward objective knowledge. Science in the traditional sense is defined as the subject of the *discovery* of the unknown, whereas art is the *interpretation* of the human phenomenon to reveal a *new perspective* not apparent or not commonly known. The difference is not in the means to find the end, but the end in itself. While science is interested in discovering new relationships about nature, art is interested in revealing the emerging social relations and establishing a relationship with nature. Human relationships are relative to time, place, and orientation; there is nothing to discover as truth since that particular new aspect of relationships changes over time. The term "democracy," for example, may be a modern term, yet its core idea has existed since the birth of humankind. Democracy has merely changed hands over time, giving rise to multiple types of democracy, including variants in which its constituents considered even the most autocratic regime to be democratic. Scientific inquiry, on the other hand, can be conducted in a vacuum with objective facts to be revealed, often with precision.

Now, if a machine is serving to reveal a new perspective of human relations, is it not an art? The means does not change the purpose. Therefore, when we study the historical authenticity of numbers to reveal a human behavior pattern using the most sophisticated econometric techniques, it is still an art. The oft-used phrase "state-of-the-art technology" honors this concept. The problem is that we have been intrigued by the power of techniques, while nontechnical social science research carries the burden of being soft and less rigorous (see Riccucci, 2010, p. 42). In fact, if we look closer into the structuration theory that was discussed earlier, we can recognize that to reveal social relations, we must use tested knowledge (structure, formal rules) to contextualize the seemingly unorganized information learned from practice. We can empirically verify tested knowledge through historical data to learn from them. At the same time, we can contextualize the "known" information to learn from an emerging situation. Therefore, the science and the art are both necessary to make sound decisions.

We can argue that decision making in social science is both a science and an art. Decision theorists dealing with public administration have largely been influenced by the rational tradition, particularly through Weber and more generally through economic behaviorists who calculate motivational principles of cause and effect to understand and predict organized actions. The focus is on individuals and their calculative actions to meet their desired ends. From that perspective, science and causal logic have priority over discovering noncausal and unknown behaviors. In contrast to the rationalist tradition, the humanist and interpretive traditions do not see individuals in a vacuum, rather as integral parts of interaction between humans (see Harmon & McSwite, 2011; Swidler, 1986; Stivers, 2009, 2011) and technology (see Latour, 2005). People are understood to have a historically effected consciousness shaped by their particular history and culture. Collectively, individual actions are shaped by the perceptions and values of others. Therefore, the individual's decision-making process is not calculative, but instead is deliberative. The focus is on the collective behavior and outcome through the series of intermediaries that mediate the new path for action. Certainly, art or the expression of the collective is the primary impetus for discovering the unknowns.

Decision making can be described as an affirmation of a prescribed act or plan by directly or indirectly influencing the plan through some form of communication. Decision making is different from opinion, in that opinion is a resource for additional information that may or may not have bearing on the final decision. In addition, seeking opinion only becomes a formality when the decision maker has enough information about the situation or there is minimal ambiguity about the predicted outcome. Having full control of the situation and the predictability of the outcome give the decision maker the "gut feeling" that it is good to move forward and forego the search for a solution through established channels. The "gut feelings" or "visions" (Sowell, 2002) are ideas not derived from any systematic process. We use our interpretive lenses to contextualize and arrive at differing visions about society. From this perspective, Sowell notes visions are precursors to theory building. However, in order to systematically authenticate a vision, one has to recreate the conditions under which it might succeed in "proving" its validity. Issues in social science, more often than not, fail to recreate such conditions because the contextual nature (time, people, and space) cannot be recreated per se. Due to this lack of a single systematic logical inquiry and the assumptions under which visions can be verified given scientific methodology, the decision-making process in social science in general is viewed less as a science and more as an art. Nonetheless, because society demands more objective proof than what could be traditionally captured through tacit knowledge (intuition), a rational person choosing to maximize his or her benefit has two options: (1) scramble for numbers to prove his or her "gut feeling," or (2) follow a predesigned plan and collect data to fit the plan, and thereby forego his or her intuition. A risk-averse rational person is more likely to settle with the latter approach. Practice-based knowledge loses to the world of *standard practice*. In other words, unorganized informal information gives way to formal standardized techniques. As Latour (1987) keenly illustrates, the issue is with a narrative along the lines of immutability and the mobility that transfers core local knowledge to universal, standardized language:[3] "The *implicit* geography of the natives is made *explicit* by geographers; the *local* knowledge of the savages becomes the *universal* knowledge of the cartographers; the fuzzy, approximate and ungrounded *beliefs* of the locals are turned into precise, certain and justified *knowledge*. To the partisans of the Great Divide, it seems that going from ethnogeography to geography is like going from childhood to adulthood, from passion to reason, from savagery to civilization, or from first degree institutions to second degree reflexion" (Latour, 1987, p. 216; emphasis in the original). In summary, although knowledge itself does not have a degree or class, one sells better than the other. The information and the knowledge that are better *represented* to the outside world often win the show. In this sense, modern ICT should give some hope to unorganized informal knowledge, which can be represented in stories, pictorials, and motion pictures.

Art of Decision Making in Public Administration

The field of public administration has traditionally leaned toward the science, relying on common public understanding of information rather than using information to construct a new, less broadly applicable notion of a situation revealed through practice. The field has traditionally leaned toward revealing what is already known and generalizable. Given that the field of public administration is more about practice and action than theory and conviction, the outcome of public administration reflecting differing human interactivities is relatively unknown. Placing more focus on human relations is essential to gaining the private understanding necessary to make decisions that will positively impact the lives of people who depend on them. Public administrators' personal understanding should take precedence over public information, particularly when public participation is low. The notion of value pluralism (Spicer, 2010) is of particular interest to a deliberative democracy. In addition, given that the outcome of decisions made in public administration cannot be known beforehand, decision making in this field is more an art than a science. The art of decision making is based on discovering the unknown through astute investigation of the human relationships and information specific to the situation. For example, if the government decides to reduce poverty in an area, the art of decision making calls for the decision maker to interact with the people who live in poverty and to know what they mean by "poor." With current and accurate information learned from and about the people and the specific situation, the administrator can arrive at a pertinent decision that incorporates knowledge not already broadly understood. Empowerment is more than just acquiring knowledge, it is about understanding how the knowledge is formed and seeking to understand the primary players involved in forming it. In this regard, knowledge specific to empowerment is tied directly to understanding the power relationship that dominates a discourse, to discovering the underlying networks of relationships among people who are pushing the envelope to describe their circumstances to other people in the world.

Once noted by Robert Behn (1996), the field of public administration can be viewed as both an art and science. Whether the field should be guided by scientific principles that can be empirically verified (facts), as maintained by Herbert Simon, or by human values, including democratic values, as espoused by Dwight Waldo, continues to dominate the discourse. In practice, however, public administrators ultimately have to deal with both facts and values or a "heterogeneity of research traditions" to execute decisions (Riccucci, 2010, p. 125). Here we can argue that the source of knowledge of the decision maker can be a good indicator of the influence of science or art in his/her decision making. Students trained to use and gain insights about policy and practice through scientific and empirically verifiable facts (data) are more likely to rely on "quantifiable" research traditions than students whose source of knowledge is permeated with human values and the philosophy of science. As I argued earlier, unless quantifiable data is contextualized given the people, time, and place to bring the data to "life," the information is reduced to becoming a perfect tool for rationalizing a preconceived end. In other words, it is the microphenomenon that is used to rationalize macrobehavior by aggregating the microlevel attributes. Michael Harmon (2006) argues that "there is no logical reason why scholars who employ micro units as their starting point for analysis cannot duly consider the systematic, relational character of social phenomena, including those macro level social collectivities they might ultimately seek to describe" (p. 43). Harmon thus captures the dilemma of using microbehavior for macro-outcome. He argues, "We are guilty of omission in our explanation of the social phenomenon if we are reporting the micro-level aggregates as 'emergent,' as opposed to merely 'additive' properties of larger social collectivities" (p. 44). We can argue that given the methodological convenience by which data can fit a predefined model, it is conceivable why many scholars find additional explanatory concepts (the missing pieces) as "noises" or mere "nuances" that are ignored or skipped over rather than explained. Technology can further reinforce the process by omitting social reality in perpetuity until a vigilant onlooker finds the scientific model to be an exercise of sheer demagogy.

Many scholars have been uneasy with the instrumental view of public administration because they argue it is incapable of encapsulating the ideas behind the practice of public administration in a pluralist society (Spicer, 2007). Michael Harmon and others have argued administrative practice is a form of moral and committed action whose proper understanding and evaluation require both a value theory and a theory of knowledge not readily inferred from the classical paradigm. Yet most of the decision-making theory in public administration, either explicitly or implicitly, has ignored the value of knowledge creation for better decision making. Referring to Karl Weick's work on generality, accuracy, and simplicity theories (1979), Harmon (2006) notes, regarding public administration theory: "Of the three criteria, generality appears to be the one valued most highly by empiricist social scientists, who then 'select' either accuracy (usually) or simplicity as their preferred 'supporting' criterion. In contrast, practitioners, because they typically deal at the level of the particular, are (if they consider theory at all) likely to value generality least among the three criteria, preferring instead the combination of simplicity and accuracy" (p. 100).

Finally, we settle the argument by answering the Socrates' dictum—all that is evil comes out of ignorance. If knowledge is the basis for removing ignorance, we might ask "what knowledge," and if there is any end to revealing? If knowledge is understood as a continuous process of self-learning in different phases of the human cycle, we must continue to invest in learning from practice to build new social relations. This is to humanize the technological power and make our actions accountable. Essentially, this is the pathway to building trust between the service providers and their citizens.

Implications of Contextualization of Information

Stories, maps, photos, and voices are contextualized information and data imbued with human values. Unlike unitary data whose validity depends on a predefined set of methodologies and techniques agreed upon by peer groups, contextualized information is a platform on which we represent others through our lenses—our interpretation, creativity, and imagination. Validity is confirmed not by an external peer group, but by the group being examined as they themselves attest to a conclusion's validity and interpretation through stories, pictures, maps, and voices. This, what we can call the *art of interpretation*, is a challenging task because the interpreter bears the burden of proof of representing others through his/her work, instead of being judged by peers applying their own standards of what is right (or what makes sense) and wrong (or what does not make sense). It is a measure of trust established between the investigator and the subject (for example, between the doctor and the patient) that can carry implications far beyond the research outcome to improve the lives of people. Therefore, the methodology used for validating contextual information and data is inherently political and subjective, hence humanlike. The investigator approaches the subject with a veil of ignorance and without a preconceived determination about the findings. Latour (1987) illustrates the difference, stating: "The unanimity between represented and constituency is like what an inspector sees of a hospital or of a prison camp when his inspection is announced in advance. What if he steps outside his itinerary and tests the solid ties that link the represented and their spokesperson?" (p. 74).

If our goal is to better serve our citizens by creating knowledge that improves decision outcomes, we must invest in knowledge building that focuses on understanding human social relations and the network that binds them. We need to humanize the data with human stories, with pictures and voices, and maps of where the people live, work, and play. This is to avoid or at least minimize systematic codification of human values. Codification of values standardizes and decontextualizes data to the extent that it reduces information to fit our perception of all that is necessary to represent others. By taking control of the technologies and methods used to represent others, we, as researchers, take responsibility to express the complex social relations that went into their constitution. Who represents whom and how they are represented are as much about power as about knowledge. Latour argues that what people represent really matters because the way people relate themselves directly links them to broader, knowledge-constitutive networks. Elements are accumulated and combined through a network of association in a knowledge cycle of capitalization. Using a narrative about Laperouse's encounter with Chinese fishermen at Sakhalin Island, Latour (1987) explains that knowledge building through representation can only become stronger if we can somehow relive the local experience:

> The first time we encounter some event, we do not know it; we start knowing something when it is at least the second time we encounter it, that is, when it is familiar to us. Someone is said to be knowledgeable when whatever happens is only one instance of other events already mastered, one member of the same family. However, this definition is too general and gives too much of an advantage to the Chinese fisherman [the local]. Not only have they seen Sakhalin twice, hundreds and even thousands of times for the more elderly. So they will always be more knowledgeable than these white, ill-shaven, capricious foreigners who arrive at dawn and leave at dusk. The foreigners will die en route, wrecked by typhoons, betrayed by guides, destroyed by some Spanish or Portuguese ship, killed by yellow fever. . . . In other words, the foreigners will always be weaker than any of the peoples, of the lands, of the climates, of the reefs, he meets around the world, always at their mercy. Those who go away from the lands in which they are born and who cross the paths of other people disappear without trace . . . the foreigner vanishes at the first encounter.

. . . If we define knowledge as familiarity with events, places and people seen many times over, then the foreigner will always be the weakest of all except if, by some extraordinary means, whatever happens to him happens at least twice. . . . *He will gain an edge only if the other navigators have found a way to bring the lands back with them in such a manner that he will see Sakhalin island, for the first time, at leisure, in his home, or in the Admiralty office, while smoking his pipe.* (Latour, 1987, p. 219–220; emphasis added)

Thus, actual information can frame our mind only through some experience. Put differently, information is constitutive from practice. What is known as the gut feeling, the way to make a decision, can only be gathered directly from locals on the ground unless it can be somehow transferred and translated to describe the actual practice.

One critical assumption underlying the rational model is that the individual decision maker is well informed about the situation, so the knowledge required to make the decision is complete. Furthermore, the model assumes that the decision maker is competent or has the required expertise (specialized knowledge) to choose the correct plan among alternatives and can arrive at a solution that is value neutral, at least from the decision maker's point of view. Such oversimplification does not question whether the decision maker has used contextual knowledge (knowledge of the situation) to make the decision; nor can it be confirmed how much, if any, of such knowledge has been applied to improve the quality of the decision. In the absence of knowledge as a resource within the rational system, it fails to validate the input, namely, information, data, and values that are used to make the decision. Rather than managing information systems (MIS), some administrators are locked in a cycle of maintaining information systems—what we may call the "MIS-syndrome." The current standards of the Network of Schools of Public Policy, Affairs, and Administration (NASPAA) once rightly included information technology as a core area of required competency. However, currently, most MPA (master of public administration) program curricula follow the traditional MIS format, resorting to business school models due to limited exposure to the literature of contextual information in public administration in the digital age.

The basic concept of contextualization is not about collecting specific types of given information which, once found, can be put together to describe the context. While one would expect one work of art to be completed by one artist if it is to give the full expression of the art, the art of decision making has to be done by a group of people who are expected to create, interpret, and explain the art to give full meaning to the context. The context cannot be rendered piecemeal, but as a fully dimensional whole illuminated by people who are connected from the beginning to the end. This requires not a structure but a professional, committed, and responsive community that "embodies a specialized skill that is capable of creating results that are both usable and pleasing to behold" (Goodsell, 1992, p. 247). Contextualization empowers public service professionals to "ask the big questions" (Behn, 1996) and motivates them to seek solutions to the complex challenges public administrators face today.

Conclusion

We teach students how not to fail by making the end known. We also show them the path to get to that end. Knowledge building, in that sense, is merely learning to follow an assigned routine without making errors. The gap between theory and practice will grow unless we bring context to classrooms. Working public administrators know that reality on the ground is more muddled than the artificial situations rational planners teach in the classroom. We can teach students how to avoid mistakes by showing the perfect plan, but to prepare them to build knowledge, we equip them to learn to make mistakes and be ready to face unknown reality. As Wildavsky so eloquently notes: "Error must be the engine of change. Without error there would be only one best way to achieve our objectives, which would themselves remain unaltered and unalterable. The original sin, after all, was to eat of the tree of knowledge as to distinguish between good and evil. However great our desire, however grand our design, we ordinary mortals can only play at being God" (Wildavsky, 1987, p. 404).

We must invest in our students' critical thinking skills and their capacity to question the ordinary and seek solutions from multiple or new perspectives. New perspectives can only emerge when we allow ourselves to imagine (visualize) what is not readily apparent to the lay person who may have experienced fewer situations and interactions from which to develop such visions. Regarding education in the new world, Cathy Davidson (2011) notes: "The brain is designed to learn, unlearn, and relearn, and nothing limits our capabilities more profoundly than our attitudes toward them. It's time to rethink everything, from our approach to school and work to how we measure progress, if we are going to meet the challenges and reap the benefits of the digital world" (Davidson, 2011, p. 20). The generation of students eager to take up today's challenges will welcome and embrace digital information, communication, and visualization tools.

The notion that social media and Web 2.0 technologies will initiate inter-activity among citizens realizing some deliberative democratic ideals may be utopian at best (see Bryer, 2011; Bryer & Zavattaro, 2011). As argued here, technology should be viewed as a relational tool rather than a deterministic tool. In other words, "information technology does not determine social practices" (Orlikowski & Robey, 1991), however, as noted, it has the potential to expand human interpretive capacity which puts the burden on individual human cognition and leadership capacities for creative action. Information technology can indeed constrain the human capacity, but, at the same time, can facilitate and enable that capacity when subjective human creativity rises above the physical limitations of the technology.

4

Leadership, Ethics, and Technology

> Leadership fails if it permits a retreat to the short run. And this retreat
> is facilitated by an uncontrolled reliance on technologies, for they over-
> stress means and neglect ends.
>
> —Philip Selznick

All ethical questions initially arise out of human agency. Technology, due to its capability to augment the mental and physical powers of human beings, stands in the role of a co-conspirator. The lure of power-enhancing capabilities makes technology an inducer of sorts, a necessary but not sufficient underpinning to many of the ethical issues we face today. Unlike any other time in modern history, having disparate datasets in one organized system allows public managers to have access to varied citizens' information on their desktop or handheld devices. The global Internet phenomenon poses a new theoretical challenge: Should we delve more deeply into citizens' information to learn *how to* solve their problems and provide the services they need? Or, should public managers contemplate *why* particular problems are persistent and focus on values that are driving the change? To answer the former (how to), we are interested in a technical tool or a quick fix—the *means* to an end. The latter requires moral attention, an understanding of the values of citizens, and an institutional commitment to improve the human condition.

As societies migrate from human habitat to the digital *infosphere*, they become increasingly dependent on digital connectedness to maintain a normal social and economic life. Unlike previous technologies that contributed to human progress, the ICT revolution has not only made societies interconnected, it has also made them much more self-aware at the local level and at the global level. Being part of the cyberworld is now considered a democratic right because it gives access to information that is considered the key to better living, better education, and an even a better personal and social life. Indeed, technology in general has had a dramatic impact in increasing the technical capacity of the human race by increasing human productivity and developing efficient means to transport resources for productive use. However, ICT's technical impact has far outpaced its impact on human progress with respect to the ethical and moral dimension. Following Catherine Davidson (2011), we must admit that we are thoroughly preoccupied with our individual selves and fail to see the broader picture of the world around us. We all suffer from some "attention blindness"—the more we concentrate, the more things we miss (Davidson, 2011, p. 4). An example in point: Millions of people today are deprived of basic necessities such as food, shelter, and education. According to the United Nations, 1.4 billion people live on less than $1.25 a day. About 25,000 people die each day of hunger, out of which 3,000 are children. The Food and Agriculture Organization of the United Nations estimated more than 925 million people suffered chronic hunger and malnutrition in 2010, and nine million children die each year before reaching the age of five. The World Food Program declared that hunger is considered the world's number one health risk—a risk greater than any communicable or noncommunicable disease known to humankind.[1] If Socrates' dictum *all that is evil comes out of ignorance* is valid, then humanity must somehow be ignorant (yet claiming to be thoroughly informed). The disconnectedness between information and action cannot be blamed on technology when our morality is not attuned to the social relations that bind us. Latour (2002) notes "to become moral and human once again, it seems we must always tear ourselves away from instrumentality . . . we must bind back the hound of technology to its cage" (p. 247). But how is technology really to be blamed when it is nothing but the product of human desire? The significance of technology in human progress in modern times leaves fundamental questions yet unanswered: What is the role of information technology in addressing the issues of social justice, freedom, and the peaceful coexistence of the human enterprise? Are these technical questions? Do they require a technical answer or a purely nontechnical humanistic response? In general, technology's ontological status should determine the type of response we might expect. Additionally, we have sided with the humanistic value-based responses to these questions. Those values can be the motivating engine to reveal solutions. We must acknowledge that as we rise up to make ourselves better protectors of humans and nature, our technologies will also reveal solutions that bind them for the same cause. Information technology can either *constrain* human creativity by routinizing human activities or it can *enable* (as interpretive lenses) new and emerging ideas to be incorporated into future actions. How it will be used depends on our own morality and judgment of good and evil. Technology can enable us to reach out to citizens so we have a better understanding of emerging values and social relations. Technology can reveal itself, ready to help. In the infosphere, we just have to use a different set of lenses to manage our own values and lead.

Institutional Value Commitment

In his seminal work on *Leadership in Administration* (1957), Philip Selznick discusses the process of institutional development for highlighting an effective leadership role in public agencies. He argues the newly formed administrative organizations are *technical instruments* for mobilizing energies and directing them toward clearly identifiable goals. Organizations, Selznick argues, are initially "governed by the related ideals of rationality and discipline. . . . It refers to an *expendable* tool, a rational instrument engineered to do a job" (1957, p. 5). He differentiates this from *institutions*, which are a "natural product of social needs and pressures—a responsive adaptive system" (p. 5). All organization must go through incremental and informal adaptive processes to earn the status of an institution. Once organizations are formed, they remain flexible until institutionalized. Institutionalization is a process by which organizations earn institutional character as they stabilize relationships both externally and internally. Externally, the organization must *secure* support by building relationships with its constituents who provide resources to run the organization and the clientele who depend on the services provided by the organization. Internally, the organizational members develop a routine informal character that binds the organization to form a *social structure*. Selznick notes that relatively enduring organizations are highly adaptable, making them natural communities. The strength of the naturalized organizations is that they develop ideologies and the sense of moral responsibility to protect group values, including a commitment to maintaining institutional integrity. Therefore, to institutionalize is to *"infuse with value* beyond the technical requirements of the task at hand" (p. 17; emphasis in the original). As organizational members become attached to the organization, they create a self-image and rise up from mere technicians to personable, committed leaders. The institutions that are infused with such values resist change, particularly ones that are coming from rational methods and predesigned goals. Just when the" organization acquires a self, a distinctive identity, it becomes an institution" (p. 21). Therefore, over time, the organization members who were initially innocent collaborators in the administration of the technical program become committed partners in building and protecting the institution. However, not all organizations show characteristics of an institution. Organizations that remain technical instruments lack distinctive competence and sense of an ideology to protect and conserve the values of the institution and the greater polity they serve (Terry, 1995). Selznick noted that to build institutions, "creative men are needed . . . who know how to transform a neutral body of men into committed polity" (1957, p. 61).

The concept of institutions has important implications in a dynamic technologically oriented society. The natural process of institution building is in constant threat by new technology deployment. The tension is so real that once institutions settle in, there is no guarantee that they will be completely destabilized by new reform measures to meet technologies demand. The resistance by the institutional leaders is miniscule compared to the large-scale deployment of management systems, such as systems devoted to enterprise resource planning (ERP), supply chain management (SCM), customer relationship management (CRM), and e-commerce operations. These systems are made to replace entire operations and the planning of several departments, and they have their own standardized mechanisms that must be matched with the rest of the business practices within any institution. Therefore, the *no-nonsense* informal institution is in a constant battle to adjust to these systems until another update and multi-million-dollar scheme is offered, promising better standardization and coordination. The evolutionary process of adapting to external constituents is now replaced with ad-hoc external shocks. The values of the institutions get convoluted to the extent that they eventually give in to the sheer rational design, where technology further standardizes the process. As discussed earlier, technology works best in a closed and standardized environment. Standardization allows the uniform application of transaction or exchange. In order for technology to realize its full potential, institutions must be first standardized to meet the demands of technology. James Thompson (1967; 2003) argues, "standardization makes possible the operation of the mediating technology over time and through space by assuring each segment of the organization that other segments are operating in compatible ways" (2003, p. 17). Standardization controls the environment by controlling uncertainty. As Antoine Mas notes: "Standardization means resolving *in advance* all the problems that might possibly impede the functioning of an organization. It is not a matter of leaving it to inspiration, ingenuity, nor even intelligence to find a solution at the moment some difficulty arises; it is rather in some way to anticipate both the difficulty and its resolution. From then on, standardization creates *impersonality*, in the sense that organization relies more on methods and instructions than on individuals" (Antoine Mas, 1949–50, quoted in Jacque Ellul, 1964, pp. 11–12; emphasis in the original).

The battle between the natural informal organization and the formal standardized machinery is real and ongoing. The resistance from committed leaders is also natural as they try to protect what they have gained over the years in terms of values they generate to both internal and external constituents. It is not surprising that most information technology projects in government die before they are launched and the success rate is less than 26% (Goldfinch, 2007, p. 917). The failure of IT projects in both the private and public sectors is accompanied by a staggering $150 billion wasted in the United States and another $140 billion in the European Union (Goldfinch, 2007, p. 918). Yet, surprisingly, the rising cost of failure doesn't appear to dampen the enthusiasm for IT investments.[2] As noted by Goldfinch (2007), "the overblown and unrealistic expectations that many have regarding information technology" continue to overshadow the failures of IT investments in the public sector (p. 918). The failures can be attributed to the lack of assessment of the institutional environment, which includes the technical (software and hardware), economic (budgetary and timeline constraints), human (skills and motivation to learn), and political (leadership and stakeholder involvement) factors that affect technology adaptation. More importantly, there is a lack of institutional commitment to take institutional responsibility for failures.

The upshot of this process is very unique: The institutions cater to the needs of maintaining the central core responsibilities, creating their own niche, and carrying the major burden of the institutions. However, the central technology systems coordinate and constantly monitor the work of the institutional members. Institutional members now have the dual responsibility of upholding institutional integrity by executing responsibly to institutional commitments while at the same time responding to system machinery that constantly monitors their actions through measurable output and outcomes. Even with the most sophisticated state-of-the-art information systems, the goal of the information technology does not change: organizing, monitoring, and controlling the existing institutional process. Institutions become techno-centric, where technology is not only a part of but is also the raison d'être of the institution. Furthermore, the institution not only depends on technology for coordination, but also for the control measures required for technology to operate under a stable environment. The elected public officer charged with producing the public's desired outcomes may step aside, ceding authority to technical experts to provide solutions to meet institutional goals (Haque, 2001; 2003). The implication of this instrumental/rational argument as explained by Thompson (1967; 2003) is: "When technology is put to use, there must be not only desired outcomes and knowledge of relevant cause/effect relationships, but also power to control the empirical resources which correspond to the variables in the logical system. A closed system of action corresponding to a closed system of logic would result in instrumental perfection in reality" (p. 18).

The concept of building value-laden natural institutions is becoming a formidable challenge for public managers who are committed to the institutional values. However, the onslaught of outsourcing from the private sector is debilitating the need to maintain leadership that nurtures *distinctive compe-tence*. This is due in part to the fact that the technological orientation of most policy mandates has blurred the ends and means distinction so that policy agendas have become elusive moving targets. Hence, the failure of technologi-cal strategies has become a scapegoat to institutional commitment and pub-lic accountability. Philip Selznick (1957) describes this lack of institutional leadership by arguing, "The retreat to technology occurs whenever a group evades its real commitments by paring its responsibilities, withdrawing be-hind a cover of technological isolation from situations that generate anxiety" (p. 79). He further elucidates that the "leadership fails if it permits a retreat to the short run. And this retreat is facilitated by an uncontrolled reliance on technologies, for they overstress means and neglect ends" (p. 82). Hence, information technologies' supremacy as a tool for resolving social issues is a necessary evil that can be at odds with the democratic values we cherish. Therefore, in a democracy, it becomes necessary to initiate a discourse about the ethics of technology in public institutions. Technological demands are often incompatible with public institutional goals and values unless they are consciously *enacted* toward meeting institutional goals. This should provide the ethical guidance to public managers in line with the democratic values, a subject that has not received adequate attention in recent years.

Technological Leadership

Philip Selznick has argued that the concept of leadership is inseparable from the concept of society. In a technology-based society, our public leaders must face the challenge of using technology to meet institutional commitments. The core of technological leadership requires a different set of lenses than what traditionally has been used. Traditionally, we have used Newtonian phi-losophy to assume societal norms can be understood by observing the sub-jects and then, based on the observation, rules and structure will be applied toward a predetermined outcome. In this case, the role of leadership will be systematic surveillance, broadly defined. This is the rational and more tra-ditional approach to leadership prevalent today. However, if we perceive so-ciety to be understood from practice, by negotiating our understanding of what society is, then the role of leadership will be constitutive or simply, in Mary Parker Follett's words, *taking orders from the situation*. Follett (1942) explains that since situations are always developing, we cannot determine the order beforehand. Therefore, the "order must be integral to situation and must be recognized as such" (p. 65). She explains, "If the situation is never sta-tionary, then the order should never be stationary . . . the situation is changing while orders are being carried out. How is the order to keep up with the situation? External orders never can, only those drawn fresh from the situation" (p. 65). The situation should guide action and not the contrary. However, since most information technologies are built for specific situations, applying them in a different situation is like trying to fit your feet into someone else's customized shoes.

Mary Follett's maxim—guidance by the law of the situation—appears as common sense to leaders whose primary emphasis is to learn on the job and then interpret the situation to avail the means at their disposal to take action. Gathering the means, including data and technology, can lead us to an end that is tantamount to a "technology of foolishness" (March, 1976). However, despite the practical advantages of learning from the situation, a rational leader seeks confidence that more data and technology are symbolic alternatives to learning from situations (Feldman & March, 1981; Haque, 2005). This is because objective artifacts (information and technology) are far more appealing notions of power than rational minds. Artifacts also simplify complexities, thereby allowing mobilization of the rules to followers (Latour, 1986). Therefore, governance from a distance becomes the default mechanism by which preconceived ideas can be set as goals, and then followers who enact the available means can be mobilized. This mechanism of leadership is much easier than being guided by the law of the situation where leaders are in constant negotiation and are learning from their constituent citizens on a daily basis.

I argue here that the technological force in our current information era provides a far more appealing platform for leadership to act from a distance without having to deal with the complexities of human interaction on the ground. The tools that are being acquired and used in the Information Age, however, are often out of context and incompatible with the actual situation. Therefore, our technological leadership may appear to have a good and powerful technological resource, but, when it comes to action, the same leaders are ill prepared to execute appropriate measures given circumstances on the ground. What to do when a new situation arises cannot be predicted, it can only be learned when our information is contextualized, accurately communicated, and adequately interpreted. The catastrophic Hurricane Katrina, the Gulf of Mexico oil spill, and the more recent April 2011 category 5 tornado in Tuscaloosa, Alabama, remind us that despite technological supremacy, leadership matters *on* the ground, not *off* the ground.[3] US Army Commander Dylan Schmorrow from of the Office of the Secretary of Defense states that the "flexibility of the human to consider as-yet-unforeseen consequences during critical decision-making, go with the gut when problem-solving under uncertainty and other such abstract reasoning behaviors built up over years of experience will not be readily replaced by a computer algorithm" (Economist, 2010, p. 18). The informal traces of information cannot be reproduced but can be reinterpreted to fit new realities. The algorithms work with quantifiable data inputs under binding rules to produce a set of known outcomes.

In the following section, it will be argued that searching situations from formal "data" (syntactic) reveals one set of outcomes that is directly associated with controlling the unknown by the systematic surveillance and monitoring of subjects. Unfortunately, decontextualized data is used to solve real-life contextual situations. Technological leadership calls for using relational tools embedded within emerging information technology tools to interpret reality as opposed to discovering them from data. Rather than getting involved in observing reality to predict the future, technology leaders take charge of informalizing the formality for understanding *values* that generate new and emerging situations.

Learning from Data

In modern-day leadership, learning from the situation is replaced with the idea of learning from data. No situation can be reconstructed piecemeal with parts of data to make the complete whole. In fact, it is impossible to recreate a situation with quantifiable data while simulating reality. This is because, first, the model of the virtual reality will be incomplete given the limited human comprehension of what the reality is. The unknowns are too many to include in a model. Even if we can gather the knowledge of the most important variables, some of the variables may have to be dropped because they are not quantifiable. Second, individuals behave differently given different times and places. As argued in my previous chapter, individuals enact technology and situations differently given the culture that has evolved out of group interaction and their own cognitive capacity. Even if we can narrow one individual's capacity to act in particular circumstances, it is impossible to narrow the capacity of action for the whole group in question. The knowledge of the situation is nonlinear, so when we try to simulate the nonlinearity, it becomes chaotic and often nonsensical. We can be 100% confident in predicting a situation given the model that has been devised, but at the same time we can also be 100% confident that the prediction will be wrong because the reality on the ground will never be the same. Statisticians claim, "torture the data long enough and they will confess to anything" (Economist, 2010, p. 6). Should leaders chase the statistical model for a predetermined outcome, or does the model chase the reality?

Data Surge and the Role of Data Mining

According to Moore's Law, the processing power and storage capacity of computer chips double or their prices halve roughly every 18 months. As the ease of owning and operating computers and handheld devices made them handy tools for average consumers, the floodgate of data opened. The decrease in the price of computers and the concurrent increase in the processing power and storage capacity allowed digital information to increase tenfold every five years. In addition, new digital devices, sensors, and gadgets are being regularly used to digitize information that was never part of any database: pictures of types of leaves, ancient archives, e-books. People are taking pictures and videos on mobile phones and storing them online over the cloud. At the rate in which mobile technology users are being enrolled, mobile phones will be the most ubiquitous technology known. According to 2011 data from the Gartner Group, a global Internet marketing research firm, there are 5.6 billion mobile phone users in the world (Gartner Group, 2011). Mobile phone users have far outpaced Internet users who themselves reached 3.03 billion in 2014 (Internet World Stats, 2014).

Although the information overflow has made data a cheap commodity, its value as a resource cannot be underestimated. Mining through mountains of data can be very useful if done methodically by using data mining techniques available to statisticians and computer software professionals.

Data mining is a widely used technique for decision making based on data. Data mining is a means of "discovering meaningful correlations, patterns and trends by sifting through large amounts of data stored in repositories, using pattern recognition technologies as well as statistical and mathematical techniques" (Gartner Group, 2012). The idea is to extract implicit, yet previously unknown, potentially useful information from a large amount of data (also called "big data"). An investigative technique and analytics serve to process the data using cloud computing processing capability. After September 11, 2001, the US government adopted the Homeland Security Act of 2002, which expressly authorized the Department of Homeland Security to "use data mining and other advanced analytical tools, in order to access, receive, and analyze data and information in furtherance of responsibilities . . . and to disseminate information acquired and analyzed by the Department, as appropriate" (Government Accountability Office, 2011, pp. 62–63). According to the statutory definition, data mining refers to "pattern-based queries, searches, or other analyses of electronic databases conducted to discover predictive patterns and anomalies indicative of criminal and terrorist activity" (Government Accountability Office, 2011, p. 61). Therefore, the primary purpose of data mining in government is to trace criminal and terrorist activity. In addition, as explained in the Federal Data Mining Reporting Act of 2007, it is also used for "detection of fraud, waste, or abuse in a Government agency or program" and ensures "the security of a Government computer system" (p. 61). By identifying anomalies in otherwise routine data (that is, snapshots of actions), data mining legally allows the detection of unwanted action or behavior that can harm public interest. Federal government data mining is widely used in the Department of Homeland Security, particularly by US Customs and Border Protection (CBP), Immigration and Customs Enforcement (ICE), and the Transportation and Security Administration (TSA). Each agency maintains its own system, namely, Automated Targeting System (ATS), maintained by CBP (US Customs and Border Protection Service), the Data Analysis and Research for Trade Transparency System (DARTTS) administered by ICE (Immigrations and Customs Enforcement), and the Freight Assessment System (FAS) administered by the TSA. Other federal departments and agencies also use data mining for creating their own knowledge management systems. Jay D. White (2007) cites GAO's 2004 survey, noting, "there is a wide spread use of data mining in the federal government" and "state and local governments are also turning to data warehousing and data-mining technologies" (p. 70). Out of 128 federal agencies surveyed, 52 are using or preparing to use data mining. White (2007) added, "199 data mining efforts were reported, 68 were planned and 131 were operational." Different agencies have adopted the data mining strategy to detect fraud and waste and to identify redundancy and duplication in agency operations. The data mining operation has been reinforced by the centralization of agencies' information technology systems connected under one enterprise resource architecture, or business resource planning (BPR), system. The centralization of the legacy systems (archived data from old systems) combined with all available data in one organized system has caused a huge surplus of data. This is a perfect mine from which "Super Crunchers" (Ayres, 2007) can use statistical techniques to generate data from data and reveal nuances that were previously unknown.

Public data is now a lucrative commodity for private investors who not only have the opportunity under the Freedom of Information Act to buy data by paying the administrative cost to retrieve them, but also can "sweep the web" for any publicly available data. These data are repackaged and customized and sold not only to private marketing groups but also to governmental agencies. The federal judiciary system is tied to two $5 billion conglomerates, Thomson and LexisNexis, which control access to all federal rulings consisting of millions of pages of public records (Malamud, 2009; Markoff, 2007).

One of the major implications of data mining in a democracy is that data is collected for one purpose but may be used, if deemed necessary, for another purpose. For example, the IRS collects income data for purposes of collecting taxes. In the process of collecting income information, other personal information such as the number of children and homes must be submitted. Then, to reduce waste and redundancy, the IRS may be legally authorized to identify citizens whose reported information indicates they are likely to be involved in fraudulent child welfare benefits. The information from the IRS then can be passed on to the Department of Health and Human Services for action. So data collected for one purpose, once inscribed digitally, is mobilized and then generalized for other purposes. There are strict privacy safeguards for citizens, including oversight by the federal chief privacy officer (CPO) and the chief information officer (CIO) who work under the guidance of the Federal Privacy Act of 2007. Nevertheless, all federal agencies and many state and local agencies can pool all data together so the respective agency can identify types of citizens or clients under the guise of cost reduction to improve agency performance. The agency can also identify who uses most services, and which constituents are politically and geographically powerful enough to make a difference in the next election. Efficiency gains and improvement in agency performance can be drivers for a more data-driven and off-the-ground approach to decision making. Therefore, whoever communicates more or seeks more services from the agency will likely leave more digital footprints for data miners to delve into for various purposes. For example, citizens who routinely receive food stamps, subsidized child care, or mental health counseling are likely to be monitored more closely than citizens who may never leave such digital footprints. The social vulnerability of a population thus becomes a trap for more data of groups that come in contact with the government for services. Consequently, a group of citizens can be under constant surveillance because they need the government to survive.

Predicting the Predictability

The data mining approach to decision making raises fundamental concerns for social equity. If decisions are primarily based on data, then whoever provides more data stimulates more decisions and orders.[4] Ian Ayres in *Super Crunchers* (2007) dwells at length on the science behind the data generating business. Data has been a valuable commodity only if it could be used for some purpose(s) (see chapter 2). When computer processing power was limited, it was almost impossible to find the needle in the haystack. Likewise, it was not cost effective to do such a search manually. Now, with supercomputer processing and storage capability over cloud computing, and with advanced statistical/mathematical algorithms out of terabytes (equivalent to 1,024 giga-bytes) and petabytes (equivalent to 1,000 terabytes) of data, many relational attributes can be derived from data. Ayres (2007) predicted that as "Quan-titative information is increasingly a commodity to be hoarded, or brought, or sold, we will increasingly find firms that start to run randomized trials on advertisements, on price, on product attributes, on employment policies . . . powerful new information on the wellsprings of human action is just wait-ing to be created" (p. 82).

As expected, the private sector has taken the task of aggregating socio-economic data to classify neighborhoods into aggregate groups. For example, the Environmental Systems Research Institute (ESRI), the map-making soft-ware giant, has created *The US Tapestry* that classifies US neighborhoods into 65 segments based on their socioeconomic and demographic composition. The segments range from "Cosmopolitan," "High Society," "Global Roots," "Scholars and Patriots," to "Traditional Living" (ESRI, 2010). The names, al-though quite provocative, can be useful classes for marketing consumer goods and making site selections based on population attributes and interests.

Whether a particular individual with given characteristics is likely to own a house, live in a particular neighborhood, or own a particular car or con-sumer item can be derived with a certain level of confidence using regres-sion algorithms. This linear decision analysis technique is now well known within the economics, business, and academic communities. However, more sophisticated techniques are now being used to create pseudo-data based on random sampling techniques that make it possible to find out the likelihood of someone not only buying the type of consumer item but also the brand, how far they are likely to drive in a given week, how many children they are likely to have, the type of school the children are likely to attend, and whether someone will get divorced within five years! Predicting the predictability is the name of the game for making best use of data. Investing in prediction makes perfect business sense because there is an enormous appetite in so-ciety to know the unknown. Citizens are powerless when their profiles are hijacked without their knowledge from their activities as regular consum-ers of goods and services they purchase from stores and online marketing sites using credit cards. Joseph Turow (2011) notes that the rhetoric of con-sumer power may not be as meaningful in the digital age, given the "rhetoric of esoteric technological and statistical knowledge that supports the prac-tice of social discrimination through profiling" (p. 3). Turow notes that the media-buying agencies spend more than $170 billion of their clients' cam-paign funds to pitch their clients' marketing goals by buying media time and space (p. 3). The impact can be so significant that it can cause widespread social distress for many consumers as their "reputation silos," created by marketing companies, bar them from entering and transacting in certain markets (Turow, 2011, p. 8).

It is not surprising that "IBM believes business intelligence will be a pillar of its growth as sensors are used to manage things from a city's traffic flow to a patient's blood flow. It has invested $12 billion in the last four years and is opening six analytics centers with 4,000 employees worldwide" (Economist, 2010, p. 6).

The idea is "if you can measure it, you can control it." What is now being called "The Age of Big Data" is having some startling impact on how we perceive data (Lohr, 2012). The industrial Internet that produces the data from people's activity is the basis on which data-driven decisions are driving productivity and generating profits. Internet usage, and more specifically, "Google hits," are primarily used as predictors of individual behavior, including what people are thinking, wishing, or expecting on a particular day or time. According to some economic forecasters, housing-related search queries, for example, have become a strong predictor of housing prices. The Federal Reserve is banking on using big data for economic predictions, and so is the United Nations through its Global Pulse program to monitor poverty using public sentiments gleaned from text messages and social network sites (Lohr, 2012). The ease and convenience of finding data is making knowledge appear to be a cheap commodity. There is not only the risk of "false discoveries" that puts hundreds of lives in peril, a lesson painfully learned from drone attacks in Afghanistan and Pakistan, but also unfair and discriminatory practices that may appear legitimate from individual data usage on the Internet, but may directly violate the rights and responsibilities of citizens when applied in practice. If data becomes the primary source of decision making, then behavioral data from people with a limited "voice" on the Internet will never be considered as decisions are made. The unheard are not inconsiderable; in fact, they may not be interacting in the digital world because, for example, they are working in the field, busily helping people on a daily basis, or perhaps on active duty in the armed forces. When knowledge is devoid of learning from value-laden practice, it is simply an exercise of biased fact-finding, an excursion to assuage intellectual curiosity. Decisions based on insufficient or unbalanced information stand little chance to improve the human condition. It is easier to monitor people through a dashboard in a corporate office than it is to govern in a democratic society.

The US government is an important player in the generation and publication of public data. The US Census Bureau has the largest publicly available electronic database of its kind in the world. It is publicly accessible online to anyone with some limited computer knowledge. By making data available about people and places at the block group or even at the block level (for limited demographics) it has opened unprecedented opportunities for social empowerment. It is also one of the most widely used data resources within the education community that provides access to data and maps about neighborhoods, schools, parks, and the public landscape in the United States. Social planners, private entrepreneurs, and marketing conglomerates search census data to understand socioeconomic characteristics of the population. Using data analytic techniques and census data, combined with other random sample data from various sources, allows powerful and accurate predictions to be made about a population group. This has become easier as the decennial census in now updated every year through the American Community Survey (ACS). This is a good thing. Rather than waiting for 10-year updates about the population, we will be able to stay abreast of changes with continuously renewed current information.

Consider, however, the profound implications of aggregating data by categorizing people by social groups and classes. When data are aggregated, they become a unique commodity—a creation that has a meaning and a life of its own. To make a structure or commodity of population data indicates a preconceived notion of the world based on a material attachment to population groups. When the aggregated data is sold as descriptive of specific groups, the buyer interprets the group in accordance with the way it was packaged and sold. For example, data about single mothers can be grouped by how many children they have, or by how many cars they have, or how many college degrees they have, or how much income they have. Each specific category can be created by identifying them and then labeling them in response to popular demand for such a category. A category can also be created arbitrarily out of sheer interest to categorize people based on prejudice. Regardless of the intent, once *inscribed*, a category becomes part of an organization or society, the language or lingo, and thence a part of the culture. Even a simple, honest coding error can cause misrepresentation of a group, which can have serious implications or concerns for the group affected by the misrepresentation. For years in a suburb in Alabama, police officers have stopped men and women with names of Asian Indian descent for minor traffic offenses but have identified them as Native Americans ("Indians") instead of Asians (Archibald, 2012). These are honest mistakes, but they can also be arbitrary when categories have to be created because of database/technology limitations.[5] It wouldn't be surprising to find such mistakes are not at all uncommon in the digital world. Categories become structures, and structures become a medium and conceptualization of actions (Giddens, 1979). Aggregating the data into a category is both constraining and enabling. It is constraining because unique attributes of individuals are reduced to a broad category of aggregates; at the same time, it is enabling because the aggregates gain mobility and become part of a language that permits the generation of that speech. Therefore, it both enables and constrains in the same way that social structures enable or constrain social action. How data is aggregated should be of important social concern for leaders in public agencies. Data should not be seen as an "item of interest" just for the efficiency gains of public organizations. Leaders should be judicious about the categories they create. There is no guarantee that a precedent set by our actions will not be misinterpreted or misrepresented or abused for private gain in ways that are counter to democratic principles.

As citizens increasingly interact with government online, they leave more digital footprints for the government. Each interaction with the government (or any other entity for that matter) is a data point. These data points are potentially valuable input for government for gaining knowledge about governmental performance limitations as well as citizen satisfaction with services. Data points can also be potentially used to gather information about citizens who visit the sites on a regular basis. The Google search engine, for example, uses data revealed in users' profiles and interests and matches such information with similar or related data on the Internet. The profiles generally also include where the user is searching from, given choices that closely match the user's location. The engine scours through all web pages over the Internet to create its own database with trillions of data points with text, numbers, characters, images, and videos that have touched the web surface. As of March 1, 2012, Google has changed its privacy policy, allowing the company to collect and share data on a user as he or she uses the company's search engine, e-mail, social networking platform, and other products. Therefore, theoretically any digital footprint created from web surfing is data anybody can capture and store in order to understand citizen/consumer behavior patterns. This data can then be analyzed through data analytics to predict future behavior. As e-government services grow at the state and local level, there are potentially important privacy ramifications that public leaders must understand and deal with thoughtfully. Matters of privacy become far more complicated when social networking sites such as Facebook and Twitter can be used to gather much more than a user's profile. The sites can reveal the users' informal interaction with friends and family, their personal views about life, politics, and society in general. Such insight might be useful in job recruiting and retention, as well as other purposes to whoever has interest, appropriate or otherwise, in that kind of information. Joseph Turow

(2011) makes a startling revelation about a company named Rapleaf, a leading data marketing company that "gleans data from [an] individual user of blogs, Internet forums, and social networks" and, according to their claims, has data on "900+ million records, 400+ million consumers, [and] 52+ billion friend connections" (p. 4). Given the rising demand for private data to understand customer expectations, data miners and investors ("data brokers") are likely to put money into broadening this capability. Undeniably, technology is both constraining and enabling. Whether its vast potential will be used to empower or abuse ultimately depends on which instincts people act upon as technologies develop. At this point, leaders can play an important role in preventing public agencies from becoming autonomous cogs of directing human enterprises without ensuring that democratic and social values have been adequately addressed in the process. "Data," as Google's chief economist explains, "are widely available; what is scarce is the ability to extract wisdom from them" (Economist, 2010, p. 4). Indeed, true wisdom for governance comes from practice-based information that is neither available in data.gov, the largest repository of government data, nor in the cookies of citizens who visit the site from time to time.

Artificial Intelligence and Neuroscience

The amount of data that is being generated with the most sophisticated techniques is overwhelming, because processing requires time and resources. I would be remiss if I did not mention the interest in using machine intelligence as opposed to human intelligence in processing the large amount of data being captured today. By automating data processing capability through intelligent recognition software that allows the detection of routines and setting them apart from the anomalies is one way of describing artificial intelligence (AI). John McCarthy, who pioneered the concept of AI in 1955, asserts that AI's goal is to "develop machines that behave as if they were intelligent" (Ertel, 2011, p. 1). It has also been intelligently described as the "study of how to make computers do things at which, at the moment, people are better" (Elaine Rich, 1983, quoted in Ertel, 2011, p. 2). The idea is to have a model of the real world in a virtual setting that can be populated with data to conduct a simulation. The machine is given precise information about each and every attribute that goes into the model. Then it can be "taught" to learn the behavior pattern of each attribute it "knows" and follow its interaction when it is in action in a virtual setting. A particular routine is expected out of this interaction. The computer can detect when the routine is broken, how it was broken, and who broke it. The computer is efficient when it comes to capturing this apparent interaction based on a linear logic model or fuzzy logic (semilinear with routine pattern). For operations that are routine, AI has become the natural method of using technology for productivity gains using semantics or theories of logic. For example, instead of humans detecting the anomalies in a product out of thousands of products in an assembly line, the machine can do the quality control with precision and high efficiency. There is no thinking involved in the process as the system runs its course through automated programs. For standardized input and output, AI is a revolutionary intelligent system created by humans.

One of the more recent uses of AI is to teach the machines how to *think* within the confines of a model and operational goals set for individual variables (players). Using complex algorithms, the machine automatically detects and extracts usable data by gauging observations in real life. The extracted information is then generalized into rules and patterns. This is a major shift from the way AI was used previously. Whereas earlier it was a machine trying to mimic human behavior by humans directly observing the pattern and teaching the machine, currently it is the machine that is given directives to observe the patterns (to pick the data) from humans or a real-life action and then letting humans know what the patterns are, so humans can learn and devise a plan of action. Given the complexity and variety of data available today, automatic pattern recognition has become the common way of creating virtual reality. So it is the supra-surveillance, or monitoring of the surveillance by machines, that will allow humans to distinguish routine from nonroutine or standards from nonstandards and to know whether to accept or reject them based on defined rules by experts or designated authority. The US government currently uses facial recognition technologies to identify terrorist suspects and detect suspicious actions by criminals. The FBI is expected to launch a nationwide service that will allow law enforcement agencies to take photographs and upload them and receive a list of mug shots ranked by how similar the features of the person are to those in the photographs sent by law enforcement officers. Heaton (2011) notes that although the privacy concerns are evident, given most photos will be taken without the knowledge of the individuals, the AI mechanism is appealing to public officials because it is supposedly an efficient way of keeping criminals off the streets.

The process used in AI is first to decontextualize reality by identifying and extracting only specific pieces of information that will make a virtual model designed to answer specific questions. Its limitation lies in its input and the logic by which the model is created. Whatever is observable and quantifiable is gathered to populate a causal logic model. What cannot be quantified and what human logic cannot grasp is either assumed given or constant, or it is simply removed from the model as "noise" (random errors). Ultimately, what the machine produces through the simulation has to be interpreted by someone. The discovery of a situation is limited to what can be observed from a distance by the machine. The final story (the narrative) about what the machine just produced (the results) has to be written or filled in by humans who will eventually have to bring their human side to the interpretation of the results. Therefore, the idea that AI will somehow make the perfect decision and that it is immune to personal bias is a myth. What is regression to statistical analysis is artificial intelligence to software programming. The results can be as bad as the data and the logic by which it operationalizes the data. Reliance on statistical evidence, rather than expert judgment, will become more common as a result of widespread data collection, from personal health to public life to private action at home and work.

Writing about AI, Wolfgang Ertel (2011) notes there are potentially infinite possibilities associated with making inferences depending on the rules of logic used by software engineers and decision analysts. What is theoretically possible to prove in mathematics is impossible to program using computer software. He argues: "Experienced mathematicians can prove theorems which are far out of reach for automated provers [artificial intelligent machines]. On the other hand, automated provers perform tens of thousands of inferences per second. A human in contrast performs maybe one inference per second. Although human experts are much slower on the object level (that is carrying out inferences), they apparently solve difficult problems faster" (p. 57).

Ertel (2011) argues that because humans use intuitive calculi, they skip the multiple steps that computers need to make simple inferences. More importantly, humans use lemmas that are tested formulas or knowledge that we already know; we "do not need to re-prove them each time" (p. 57). Ertel asserts that an attempt to formalize intuition is problematic because it is difficult to relay knowledge using formal language and then interpret the same. Therefore, as far as intuition is concerned "we cannot program this knowledge or integrate it into calculi in the form of heuristics" (p. 57). The limitation of the algorithms is not the data (in fact there is too much of it) but the machines' ability to capture the traces of data that reside in real-life situations.

Many studies in mainstream neuroscience suggest that much of what we do is hardwired. It is tempting to believe that further research will eventually demonstrate that physical properties of the brain fully control the human mind, yet one of the most important discoveries in neuroscience in recent years points out that humans have a built-in interpretive capacity in the left side of the brain that builds a narrative in each of us. Neuroscientist and psychologist Michael Gazzaniga (2011a, 2011b) argues in his Gifford lecture there is enough evidence to conclude that human behavior is not fully predetermined. He argues that a sense of responsibility, for instance, derives not from within a single brain, but from social interaction. He maintains:

> We humans seem to prefer black and white answers to questions, binary choices, all or nothing, all nature or all nurture, all determined or all random. . . . I will maintain that the mind which is somehow generated by the physical process of the brain, constrains the brain. Just as political norms of governance emerge from the individuals who give rise to them and ultimately control them, the emergent mind constrains our brain. . . . In a time when we all think we can agree that causal focus are the only way to understand our physical world, are we not in need of a new frame of thinking to describe the interaction and mutual dependence of the physical with the mental? . . . Even with all the knowledge of physics, chemistry, biology, psychology, and all the rest, when the moving parts are viewed as a dynamic system, there is an undeniable reality. We are responsible agents. (Gazzaniga, 2011a, pp. 4–6)

Gazzaniga notes that after years of research, he and his colleagues have found that the left side of the brain (responsible for controlling speech) has something called the interpreter that appears disconnected (or blocked) from the right side of the brain that controls vision, yet surprisingly the left can interpret what the right side of the brain may be doing given what it learns from the surroundings. He argues, "Our dispositions, quick emotional reactions, and learned behavior are all fodder for the interpreter to observe. The interpreter finds causes and builds our story, our sense of self. It asks, for example, 'Who is in charge?' and in the end concludes, 'Well, it looks like I am'" (Gazzaniga, 2011b, p. 938). Gazzaniga maintains that we are constantly interpreting the world given what we attach. There is nothing called free will in society; we are bound in a social milieu. There are no automated directions that will allow us to reach the fundamental truth about society. However, in the same vein as Cathy Davidson (2011), Gazzaniga argues we are responsible for our own actions given "what we learn, unlearn and relearn" through our interpreter. He states, "Personal responsibility is not to be found in the brain any more than traffic can be understood by knowing about everything inside a car" (p. 938). Gazzaniga believes all network systems, for their proper functioning and survival, must have an enabling and constraining mechanism built into the system. He states, "Human society is the same. Responsibility is a rule established by people. Researchers might study the mechanistic ways of the brain–mind interface forever, with each year yielding more insights. Yet none of their research will threaten the central value of human life. It is because we have a contract within our social milieu that we are responsible for our actions" (Gazzaniga, 2011b, p. 938).

The US Defense Advanced Research Projects Agency (DARPA) in 2011 solicited research proposals for understanding how narratives influence the neurobiological, chemical, and psychological instincts that mobilize human actions. The primary goal of the proposed research agenda was to "take narratives and make them quantitatively analyzable in a rigorous, transparent and repeatable fashion" and "identify and develop narrative analysis tools that best establish a framework for the scientific study of the psychological and neurobiological impact of stories on people" (General Service Administration, 2011, p. 7). As per the traditional scientific inquiry, the idea behind this multi-million-dollar research initiative was to decontextualize the stories to find out the neurobiological cause that triggers actions.

What is critical to note is that once the reality is decontextualized for finding the cause, it has to be reconstructed and recontextualized by humans in order for them to apply the concept in real life. Here is where the role of leadership becomes critical. When all is said and done and the scientists have produced their findings, for example, in finding the neurological, psychosocial, or chemical causes that stimulate the brain to mobilize individuals for action, the responsibility falls on the leaders, specifically the public administrators, who are given the responsibility of taking the findings (causes and effects) and implementing them in real-life situations. The expectation then is as follows: If the facts are right in the "laboratory," we can find similar situations and look for causes that can lead us to the end result as expected from the "laboratory" findings. Even if it were imperfect, from the real-life test we would know what did not work to achieve the predetermined outcome. The determination is that what the scientific community (thinkers) was able to deconstruct, the public administrators (doers) should be able to reconstruct. The idea of course is that the nonquantifiable social pieces that were excluded from the laboratory model and assumed constant (having no significant role in the model) are now alive and well in the real-life model. In other words, what was excluded in the laboratory model now must be brought into real life if one is to make sense of the causes that are expected to have an impact on the outcome. The leaders' task then is to use their "interpreter" to make appropriate use of new scientific directives if they are to improve the human condition. Human art begins where science has left off! Contrary to the traditional belief, leadership responsibility increases rather than decreases in a rational, technology-driven society. Therefore, in order to execute responsibly, broad discretionary authority for public administrators is critical to maintain the balance between what is expected from the rational design and what can be done practically to apply that to real-life situations. Many contemporary failures in government are due to the overzealous expectation from technological imperatives that are logically or rationally driven and yet practically disconnected from reality.

At the height of rationalism during the French Revolution (1789–99), English philosopher and statesman Edmund Burke foresaw the critical role public administrators played to preserve the democratic values and traditions of enlightened Europe. He declared to public administrators: "The laws reach but a very little way. Constitute government how you please, infinitely the greater part of it must depend on the exercise of the powers which are left at large to the prudence and uprightness of ministers of state. Even all the use and potency of the laws depends upon them. Without them, your commonwealth is no better than a scheme upon paper; and not a living, active, effective constitution" (Burke, 1901, in Haque, 2004, p. 705).

Leadership Praxis

Leadership is about taking responsibility. Responsibility increases with technology in the mix because now we have added responsibility to ensure that the technology being used serves the primary purpose of responsible action. Technology adds to the convenience of governing from a distance, but it increases the responsibility for the consequences of using the tools for public service. Leadership in a technological environment is far more challenging in a democracy than it would appear in depictions in popular media. Scholars in the field of public administration have been highly critical of rationalist ideas in public administration because they see that, due to the limits of rationality, many policy directives will partially or never materialize to improve the human condition but rather can be instruments of administrative abuse and undemocratic use of administrative power (see Adams & Balfour, 1998; Adams & Balfour, 2009; Spicer, 1995; Haque & Spicer, 1997). It is not that the use of technology is a bad thing; rather, it is the use of abstract ideas (in policy planning) that, when applied via technology, gets its legitimacy as it becomes the *apolitical enabler of action*. Learning from past administrative praxis is indeed lost in this process. The technology can then become the perfect guise under which any administrative abuse can be legitimized or any ethical violation can be cloaked. Thus in a technologically driven rational society, leadership can become immune to responsible public action.

The source of leadership can only come from relationships between humans. More specifically, to improve the human condition, leadership should invest most in the capacity to create long-term relationships between the leaders and their constituents (Harmon & McSwite, 2011) and less in artifacts including abstract designs. Technology should be used to build human relationships, not to reinforce predetermined orders or situations that are detrimental to relationships. Because relationship is a social value—found in neighborhoods, on street corners, in public hospitals, and in the peripheries of urban suburbia or a rural town—leaders must use their social skills and technology skills to understand and nurture the values that are key to building relationships. There is no leadership app or government app that can help us understand values concerning relationships from a distance. Relationships require active participation, interpretation, and learning as they build the network of relations. As James March in his recent interview reminded us: "Being a leader involves being embedded in a network of relations and expectations that considerably reduce one's autonomy. By and large, of course, society does not want 'autonomous' leaders. They are dangerous. For every leader who pursues an autonomous direction that benefits society, there are many more who pursue autonomous directions that harm society" (Podolny, 2011, p. 504).

Information gathered from the public must be citizen-centric, specific to addressing the needs of the citizen. We know more about the cause and effect than we know about the subjects. The information inequity puts the government at odds with democratic principles. Government can easily collect more information from people who have more contacts with governments. Meanwhile, people who have more contacts (active citizens) are the ones who use or need the government the most. Therefore, active citizens are the likely government targets for more information gathering, and they are also the ones whose data will increasingly be used to illustrate the model of an omniscient government. Even with the information overload in the Internet era, the flow of information between a government and its citizens remains one-way traffic; the government knows more about the citizens than the citizens know about the government. The situation is further skewed because some citizens do not have the means to access government information. Clearly, the information asymmetry is at odds with democratic values. As more and more information is collected, more and more categories are created to organize it. Consequently, more policy prescriptions are directed toward people whose data is available. Even if we are informed with the data, we may not know how to interpret them in all their associations. For example, the leaders might collect enough detail to refine the definition of poverty, and they may devise sophisticated indexes of poverty thresholds, yet still they may know very little about what it means to be poor. Knowledge about a subject must form through information that is constitutive. Data accumulation about a subject can become a pedagogical exercise that may be theoretically relevant yet practically irrelevant given the knowledge that is formed to develop the theory is only on paper. Data reduces complex practices into generalizable norms that are "*immutable and combinable mobiles*" (Latour, 1987, p. 227). As Latour notes, "they all take the shape of a flat surface of paper that can be archived, pinned on a wall and combined with others; they all help to reverse the balance of forces between those who master and those who are mastered" (Latour, 1987, p. 227). Data is a commodity that is bought and sold, used and reused to understand the reality of others. To an investigator or a policy analyst it is a virtual reality until he/she accidentally encounters a situation that reminds them of what that *reality* is.

Data about people does not help build relationships, but it can help organize people for systematic monitoring and surveillance (Giddens, 1987). The true definition of leadership praxis therefore is to correct the imbalance created by systematic, one-sided data collection by exchanging it for a two-sided social relationship. Learning evolves out of social exchange as one builds relationships to interpret the life of others through their lenses. Latour (1986) noted: "An asymmetry is created because we create a space and a time in which we place the other cultures, but we do not do that same. For instance, we map their land, but they have no maps either of their land or ours; we list their past, but they do not; we build written calendars but they do not" (Fabian 1983, quoted in Latour, 1986, p. 16). The apparent disconnect between managers and the citizens they serve cannot be resolved with data transfer. For an equitable and just society, value transfer is more critical. On the one hand, there are large concerns about big corporations and multi-million-dollar technological and scientific research communities' influence in controlling the application of technology. At the same time, consumers are being targeted and manipulated by technocrats to get fixated only on new technology for better techniques of advancing society. Information technology has been used as a medium by which a class of society benefits out of a sheer improvisation of technological supremacy. In the name of transparency and citizen empowerment, the federal government has also embraced the idea of extracting policy ideas from data mining. The major part of data.gov is devoted to the business of data mining and data monitoring.

Conclusion

Organizational leadership must mediate technologies to develop and nurture distinctive competencies that are geared toward protecting democratic values. As we are being molded by information, so is our morality. All new technologies are being forced to become routine, and new events once un-heard of suddenly become quite common (just as we become immune to deaths when we see scores of them happening on a daily basis due to wars and natural tragedies). Our morality is often judged relative to the scope of the problem. The more unusual and bigger the evil, the greater are the Tweets and Facebook Likes. Information technology brings such stories to us to in-form our passions and test our morality. In this sense, morality and tech-nology shape our ontological nature of the world (Latour, 2002). Therefore, with the qualitative improvement of our technology gadgets, our "quality of morality" must also be improved through self-reflection. Our leaders must be distinctively competent in reassembling morality and technology by learn-ing from context and the historicity of events.

5

The End of Surveillance

The most effective control over administrative processes begins with acknowledging the impossibility of closing the gap between idea and action. . . . Can we explore the possibility of another kind of control, one that takes its orders from the situation? Can we pay attention to actions that enable the visualization and expression of the contingent, the arbitrary, and the particular?

—Camilla Stivers

My final chapter is not intended to settle a debate, but rather to open one. It is the discussion of the beginning of an end to surveillance in the Information Age. Information must be understood in human terms, not in terms driven by technology. How do we utilize technology to be informed—to shape the human mind—through practice? How do we use information technology as a mirror of information about ourselves, a mirror that will let us see our own social order from local constitutive practices? How can our practices and beliefs show us how to interpret the world in ways that could make our practice a shared experience rather than a competition to survive in the Hobbesian jungle? Information, after all, is constituted from practice. Technology is a communicative agent that allows a practice to be routinized in ways that reinforce the existing social order or control the incongruity within the social order. The impending task is to explain the need for public administrators to develop an enhanced set of skills for using information technology in order to take advantage of the practice-based and potentially democracy-promoting features of the technology.

Information and Practice: The Missing Link

In a rational world, information gathering is used to meet the purposeful ends of society. Ends can be programs, systems, or organizations that have been created to serve a given purpose in society. Once the ends are determined (presumably through a democratic process), information is strategically and selectively gathered in order to ensure that the desired ends are being met. The selected information is the measurable data that indicates the desired outcome. The measurable inputs also ensure that the people responsible for maintaining the ends (programs, and such) are also accountable given the expectations of the program. Hence, as described here, a system is born and a process is in order with the help of information to *administer* (organize and monitor) a desired end (program). Information technology comes in very handy in processing the collected information, but also in maintaining the integrity of the system by avoiding human bias entering the system (think about an assembly line, for example). Therefore, we can argue that in a rational society information is primarily used to *administer*—organize, process, and monitor—a system created by society, and technology reinforces the administration's capability. The administrative focus is meant to ask how to solve a problem. It does not beg the question, why solve it?

How does the new administrative system survive given other systems that are also generated by society on a continuous basis? The survival of any new system will depend on how well the measurable numbers from the information collected reflect the successes of the system. Failure to produce positive signs from the system will likely diminish support for such a system. In order to protect the system, people working for the system will generally tend to show more successes than failures. In other words, the measurable information will be inherently biased toward showing the success of the system rather than the failures. Here, information technology plays an important role. It will be used to gather more information (some less useful than others) and produce successes more "quickly." Therefore, *administration* in the modern Information Age is expected to exemplify two important characteristics: (1) It will always seek more information than it needs to prove its worth; (2) it will advance technology to produce successes more quickly. *Learning from practice* to ask the question, *why do it*, is *not* a characteristic built into the modern administrative system. Using the discussion here, we can argue there is a direct connection between information and administration, but there is no apparent connection between information and practice. In other words, we will use information for administering systems, but we are unlikely to use information to learn from practice. Clearly, the growing disparity between information and practice is due to our increasing dependency on information that is used for *administering and surveillance, rather than learning from situations—practice*. A typical case in point:

Each year, thousands of juveniles enter detention and correctional facilities and treatment programs throughout the United States. Many at-risk youth spend significant time in one or more of a variety of public and private detention facilities and residential treatment homes. All but a few of these young people will eventually return to the community. Most that do not make it remain on the streets, in detention centers, or wandering from jobs to "jabs" to make a life or make a living. In the United States there are many federal- and state-funded programs that focus on how to keep youth away from idleness on the streets or in detention centers and, instead, return them to the communities where they can be a productive part of society. The programs traditionally use the Intensive Aftercare Program (IAP) model based on data-driven research that claims that a highly structured and enhanced transition from confinement to the community would benefit youthful offenders in areas such as family and peer relations, education, jobs, substance abuse, mental health, and recidivism without negatively affecting the community.

I had the privilege of being associated with a state-funded juvenile aftercare program in Alabama for almost a decade. As an evaluator and program advisor, I had the opportunity to observe how data as information/knowledge input had direct implications on the expected outcome and the values that were imposed upon the observers through the data-driven approach to decision making. The program titled Return to Aftercare Program (RAP) was designed for first-time juvenile offenders (10–17 years of age) who are at risk of recommitting an offense due to their mental health and behavioral issues. By addressing mental health needs through medical and social services, and by preventing bad behavior through graduated sanctions as well as incentives for good behavior, the program reduced juvenile delinquency. The primary goal of the program, based on a widely used model by the Office of Juvenile Justice and Delinquency Prevention (OJJDP), was to infuse positive behavior among at-risk youth by exposing them to community-based good behavior practices and monitoring their daily activities, including behavior patterns, academic performance, mental health, and family support services. Various types of individual client data were collected at different phases of each client's involvement in the program. The data input ranged from personal data to family history, personal past medical history, mental health, psychosocial needs assessment data, and academic performance records. While in the program, the clients were extensively monitored through either electronic monitoring devices and/or frequent house visits by probation officers and case managers. The clients were required to report to the family court judge on a regular basis to discuss progress in the program. Based on behavior and school performance as reported through probation officers to a judge, a juvenile would either graduate after spending time in the program or get new sanctions and/or new placements.

The data was the primary vehicle by which program managers and state officials could evaluate and monitor the success of the program. All data was organized in a centralized database managed by case managers and probation officers. The data had to be measurable, so it had to be either directly quantifiable or coded to make the analyses. Like most public programs, the success of the Return to Aftercare Program was determined by a measurable outcome, specifically how many juveniles were able to stay out of trouble while they were enrolled in the program. A client showing positive behavior for a sustained period of time would graduate from the program when a judge deemed this appropriate. According to the goals set for the program, it was highly successful. With more than 75% of juveniles logging no recommitments at the six-month review point, the program was regarded as a model for similar programs across the state and the Southeast. The graduation rate was between 50% and 60% (not including cases that were closed due to the client's absence without leave or other contingencies) over the nine-year period. The data provided, per its formal institutional design, an enormous amount of information about how to graduate a first-time juvenile offender successfully. Specifically, the data would tell us the characteristics of the participants in the program who are likely to succeed and the ones who might not make it to the end. Together, the programmatic performance data and individual behavioral traits could develop predictability models of graduation/nongraduation. However, the data provided almost no information about why so many juveniles who enrolled in the program were unsuccessful or decided to leave the program before they graduated. We knew quite a bit about what it took for someone to graduate, but we knew very little about *why* so many decided not to return to the community. According to the program design, the accountability to the program goals was measured by the successful graduation rate while it excluded any other contending factors.

As service providers and researchers, we were focused on the efficiency of a successful outcome based on data generated for the project. Contextual experience gathered from local practice was not translated into the project. In fact, we always learned something from our local experiences, but because experiential learning was secondary to data originally designed to be incorporated into the existing programmatic structure, it was either ignored or never formally considered to be part of the program's outcome. Our interest was primarily in meeting the goals of the program and focusing on data about what leads to success. What remained relatively unknown throughout the project was *why* juveniles were unsuccessful in the program. The reasons for lack of success remained unknown in the measurable data not only because capturing the reasons was not initially part of the formal design, but also because all behaviors cannot be quantified in ways that can fully uncover each individual's motivation. Such nuances must be deduced in consideration of the individual's situation. In other words, we could conjecture and make hypotheses given the data available from the juveniles' activity (number of offenses and behavior history), but the real-life factual evidence of what led juveniles to quit remained hidden, the outcomes having been reduced to "success" or "failure." Again, the question of why they succeeded or failed remains unanswered because it cannot be measured, it can only be observed and narrated through a series of situations.

Learning from situations (contexts) is a slow process and requires human involvement much more than the fast, data-driven approach to understanding decision outcomes. Technology, as used traditionally, provided the impetus to abandon the slower and adaptive approach of learning from practice and to adopt the efficient, measurable, data-driven approach toward decision making. In a society enthralled with technology-based information, situations are not captured in routine data collection efforts, they are left for the experts to assume or tease out of objective data analyses generated to meet the goals of the program. In our study, the situations could be gleaned from comments occasionally left by case managers. For example, when one juvenile was reported AWOL after two weeks, a manager's annotation revealed that the juvenile did not return to the program after being "kicked out from home by step-mother." Another juvenile never returned to the program after, due to financial problems, she stopped taking the medication she needed for mood behavior. Coming across some of these comments, I realized, given the data, we were indeed meeting the goals of the program, but we were not meeting the goals of the youths or society to make a difference in their lives. The ends—success or failure—were already defined in the project. All that remained for the service providers was to fill in the blanks about who qualified and who did not. Therefore, the data generated for the project were meeting the demands of the project, but not the demands of society. Since the end was already defined for the project, the value-laden information gathered from local practices had no chance to be included in a database or into an information system. If noticed at all, those aspects remained only in the minds of the observers. If we are to make a difference in the lives of people, the end must emerge from practice, and our role as practitioners must be to observe and translate that observation into our programs. Camilla Stivers, in her work *Governance in Dark Times* (2008), reminds us: "Ends must emerge and develop through experience. . . . It is in the experience of the work itself that we are able to create its meaning. The only pragmatic recommendation to make in the idea of creating value *for* stakeholders is that we create value *with* them. Values are not injected like a vitamin shot into shared experience, rather they grow out of it" (pp. 128–129, emphasis in the original).

If the goal of the program was to find out *why* many juveniles decided to quit the program, it would have been wiser to invest in narrating the network of relationships of these young adults than to collect each and every bit of data about the program's activity. Needless to mention, the bits of measurable data collected in parts must be collated to make sense of the big picture of what influences individual action. Measurable data driven by institutional goals can only provide partial answers without giving much clue about the networks of relationships that indicate societal goals. Ann Rawls, citing Garfinkel's (2008) work, notes that the power of inference from bits of data is hampered when relationships are rationalized (abstractly created) as opposed to being communal (relationships as understood from practice). She adds, "the ability to make mutually intelligible inferences depends on having a high degree of congruence on primary [direct] and secondary [reflective] information—and that in turn requires the constitutive features of a communicative net—which rationalizing a system interferes with" (Garfinkel, 2008, p. 73). The practical information provided through relational narratives also gives us powerful clues about the social relationships and social order. Clearly, data devoid of values is problematic when it comes to learning from local practices and understanding the social order. It is, however, efficient in producing results sought and limited by the formal constraints built into the design of the program.

The constraint imposed through the design of the project discussed here is primarily a function of technology's efficiency in measurement and organization of the inputs (in the RAP program, participant attributes and activities) and measurement of the output/deliverables (success or failure). This functional design greatly influences how technology is enacted in social science projects in general. It is not at all inconceivable that technology can be used to *inform* the practice, as opposed to *administering* the practice. However, we tend to avoid taking risks by following the path where the end is sharply defined in binary objectives: success or failure. Because we are expected to show success, not failure, we invest more energy in *managing* success. Our emphasis on the success of the program was clearly more reflected in our focus on the determinants of success than on "other" factors that were mere anomalies as far as our goals were concerned. We clearly suffered from "attention blindness" syndrome and missed "the gorilla in the room" (see Davidson, 2011). We paid too much attention to data that defined the success factors and fit well with our predesigned model, but remained blind to data that could indicate the causes of the failures. We naturally eliminated the "noises" from the model that did not document success.[1] Most often "noises" are unstructured and cannot be measured or quantified but are understood as constitutive practices that can only be explained by situations. Following the discussion about the Return to Aftercare Program discussed above, we can now extend the logic behind the researchers' intrigue with measurable outcomes: once we find "what causes good behavior" we can move ahead in a single-minded search for those behavioral traits in juveniles who would be considered potential successful cases. Our findings might also prompt neuroscientists to analyze the brains of potentially successful cases to investigate what causes such good behavior. Consequently, what happens to cases that are unsuccessful is that they are left aside and considered someone else's problem. But should we run away from the unpredictability and the "noises" that do not fit our preconceived plan? Information is not what is already known; rather, it is "something that is different, unexpected. It is the anomaly. That is what makes information" (Garfinkel, 2008, p. 36). How can we be *informed* when we choose to "know" only what is expected from retrospective accounts rather than to seek what is not yet known?

Information gathering is misleading when the gathering is based on the premise that it will be used for a purpose already defined. Garfinkel (1967) notes that information is most useful and relevant when discovered from practice, not from retrospective accounts. Do we collect the information and then ask, "Is the information worth the cost?" or do we ask, "Will it have been worth the cost?" (Garfinkel, 1967, p. 193). We cannot go back and collect the information that we have ignored all along, assuming it unimportant given the task at hand. By the time we have finished collecting the information, the task is no longer the same. New issues will have come up that call for the collection of new sets of information that were not part of the variables included in the previous model. Structured data models will always run into problems when one is trying to capture situational information. An unstructured and open model that has the ability to capture informal behavior to give investigators ideas about social values will always give far more insight into situations. In the final analysis, the limitations of a project in terms of budget and time will circumscribe what sort of valuable information will come out of the project. This leads us to conclude, with Garfinkel, that the motivation to collect "core" information from the project becomes secondary to "good reporting performance" (Garfinkel, 1967, p. 194). He further clarifies: "An administrator with an eye to the budgeted costs of his reporting procedures is apt to prefer to minimize the burden of the present costs and to favor short-term peak load operations when the investigator has decided his needs in a formulated project" (p. 194).

Given our perception of how technology can be used, we tend to rely on the constraining factors of information technology to *inform* about the situation more than about the enabling factors. By putting faith in objective data, we constrain the technology's ability to translate information about the contextual, local narratives and the relational networks within situations. We are clearly motivated to see successful results exemplified in measurable outcomes. The problem is that we would thereby know more about experiences where we did well than experiences where we did poorly. Denrell and March (2001) coined the term "hot-stove effect" to depict this phenomenon. They extrapolated it from Mark Twain's wisdom: "If a cat ever jumps on a hot stove, he will never jump on a hot stove again. And that's good. But he will also never jump on a cold stove again—and that may not be good" (James March quoted in Coutu, 2006, p. 86). In other words, the alternatives with positive outcomes are likely to be reproduced and refined as more information becomes available. However, the outcomes that initially did poorly will never be tried again, even though their potential has not been tested with additional information that could correct the errors. The potential to improve an alternative outcome is stalled as a decision outcome is influenced by how quickly the positive results can be discovered. That path is not only risky but also dangerous, because it has not been chosen from all possible alternatives (negative and positive) but rather is the one that initially produced the most number of positive results. To search for gold, do miners look for pure gold mines (there is no such thing as pure gold mines) or do they look for related metallic ores—for the silver, copper, or uranium that are primary indicators of gold? To find any gold, we must pay attention to the metallic ores, the "noises" in the system. The domain of information can be at least as valuable as any specific bit of information chosen from within the domain for better understanding of any situation. Denrell and March (2011) point to the dangers of a mindset where we value success only for the sake of knowledge building: "By reproducing successful alternatives fast learners quickly eliminate poor performing alternatives in favor of better ones. The very quickness that makes such a process particularly effective, however also makes it biased against risky and new alternatives. By continuing to reproduce failure, slow learners are less likely than their quicker cousins to suffer from the hot stove effect" (Denrell & March, 2001, p. 532).

Is it enough to end our research once our programmed technology disgorges the evidence of our success? No, we should persist and implement technology designed to enact our inquisitiveness to be *informed* from the realm of unknowns, even if it is painstakingly slow and wrung from *failures* that do not meet our expectations. Technology, in this sense, is used to connect to the broader *social relations* between alternative choices available to us to make decisions. When we have technology, we pretend to be empowered by it in the sense of being armed to find the "truth" quickly. History teaches us that learning is most effective when it is accomplished through a slow adaptive process because it allows failures to be corrected and tested from experience. Whereas a novice gold miner searches for "quick gold," an experienced miner searches the *broader relations* of metallic objects that can tell where to find not only gold, but also other valuable metals, even in the most unexpected places.

Practice-Administration Dichotomy

Administration is to organize—to constrain information—and practice is to translate information into communicable action. Practice is flow concept; it is constitutive, as information is used in building relationships. Administration is a static concept; information needs to be held, monitored, and organized for creating a product called service. Students of public administration learn about the politics of administration dichotomy. In the Information Age, they must understand and engage in the practice-administration dichotomy. There is no methodology for organizing people without breaking the continuity of the social design. Whereas social design is constitutive of human interaction, its survival depends on the continuity of shared practices. In order to break the continuity to introduce objective standards of organizing, we must break the society into individuals. While we attempt to do this, we also break societal practices into individual parts based on individual attributes and practices. While we move deeper into finding individual motivation or behavior, we move away from understanding societal practices and collective behavior. Although individual analysis is inherently deductive, examining cause and effect relationships and collective behavior is inherently constitutive with unpredictable ebb and flow.

Communication-enabled technologies not only inform practice, they also change our perceptions about people and societies known and unknown. Every act of communication is an act of translation (Rabassa, 2005). Technology as a communicative tool participates in that translation process by informing the choices we make to influence others. When information is provided so as to limit the choices of others, clearly less information is translated, if not distorted. Information is not what *is expected* but what *is unexpected or unknown*, so anything that is predictable constrains the ability to learn from the realm of unknown possibilities. Therefore, when technology is used to limit one's choices of predictable information, it not only limits the amount of information but also changes relations of power.

As citizens and administrators using technology in our daily lives, we participate in the translation of information. We must be careful what we transmit (through translation) to the society using the technology. We must conceptualize technology as a communicative tool that informs how individual values are exemplified in practice. Rather than asking the question *how do we use technology* (means to an end)? we must ask *why do we use technology*? The difference in the question that we ask fundamentally changes the purpose and the value of public service. It changes the ontological status of technology. As stated earlier, when we ask the question about how to do the task, we reduce the work to finding the means to our desired end. When we ask why we are doing it, we put the end first before we look for the available tools that can help us achieve our purposeful end (see figure 1 in the introduction). The question of why we are doing it is valuable because it is an emerging question; it is not absolute. It changes with the situation; it requires us to ponder, to relate to otherness, to reflect and to learn from practice.[2] How to do the task becomes critical when we know we are doing it. Technology can help us in achieving both—thinking and doing. As argued earlier, the quintessential value of public service emerges when we are able to reconcile thinking and doing by focusing on developing ontological solutions to both sides of information at the same time.

The indirect consequences of technological impact are subtle yet powerful, having enduring implications in the lives of people and their social norms.[3] Indeed, a catastrophic event can be more damaging psychologically and socially, and wider in scope and time, to people who are indirectly affected than to people who received the direct shock. This is because people who are indirectly affected by catastrophe receive the information through translation. The means of translation and the translator's worldview are both fundamental to what information can be deduced from the translation. Therefore, the human mind is in full control of the destiny it chooses from the continuous translations from the social milieu, from the source of information of ideas and information (correct or incorrect, good or bad, applicable or inappropriate) from which the brain develops (See Gazzaniga, 2011a; 2011b).

How can the human mind use technology to open choices for others? It is the *interconnectedness* attribute of technology and not the *integration* attribute that is the most powerful liberation force to nurture democratic ideals in the Information Age. Because power relations are embedded in the translation of information, we have to focus on the information technology and the types of information it creates to inform and organize individual action. In public administration generally, technology traditionally has been appealing to bureaucrats because it can integrate information—to administer, monitor, and control the affairs of government. The intended purpose is as clear as Woodrow Wilson (1887) famously declared, "it is getting to be harder to *run* a constitution than to frame one" (p. 200). Wilson was arguing to systematize administration under a "branch of science" (p. 198). In his view during the Progressive Era (1880s–1920s), it was becoming increasingly difficult for government to manage and control the uncertainty and the complexity in the affairs of government. He argued in favor of gathering information to define the objective end of government and then organize and manage that information toward the defined end. Wilson was focused on the *integration of information* side of bureaucracy. This, he thought, would be a practical means to "discover what the government can properly and successfully do" and do it "at the least possible cost, either of money or of energy" (p. 197). Clearly Wilson was influenced by scientific logic and would probably have supported using information technology, had it been available then, to control and manage the bureaucracy. However, Wilson's prescription for the control of bureaucracy was not just due to his faith in science; his decision was significantly influenced by the context and culture of the time to protect democratic ideals. During a time when we propounded the science of administration, the spoils system had insidiously undermined bureaucracy. Wilson was convinced that objective facts and the standardization of information could control unintended bureaucratic behavior and make bureaucrats more businesslike. This in no way implies his unbridled support for efficient administrative apparatus at the expense of democratic principles. Undoubtedly, he would support standardizing the administrative machinery to meet the ends designed by the elected officials—a perfect (standard) machinery to be used for a known (predictable) outcome. When information is schematized (noises removed), it is far easier to manipulate that information to produce results that are representative samples of success. The result of success is produced, while the failures that "fell through the cracks" are ignored. In other words, we have developed a perfect system to show how to write a successful story through quantifiable information and ignore why it failed. The street-level bureaucrats never had the chance to tell their side of the story as to why the program succeeded to some extent but largely failed because the system was not designed to capture the details. How perfect a system, how perfect the plan, and how precise the known outcome!

Public administrators are not immune to being part of the problem when the street-level bureaucrats (informal groups) are outside the formal organization's direct sphere of influence. Without value commitment by the bureaucratic leadership, institutions will fall into "group think" behavior where distorted judgments can become true facts (Asch, 1951). Undeniably, the collective construction of rationalization based on technological schemes to generate quantifiable outcomes can become dangerous when it signals moral beliefs that are antithetical to democratic values. The work of Irving Janis (1972) should be instructive in this regard.

The implication of this argument discussed so far is that public administrators have fallen into the administering, monitoring, and maintenance of the *formal* decision apparatus that teaches *how to* produce results. We also teach students of public administration about techniques that will produce faster and better results. The reliance on technique is unabated with institutional reforms calling for approaches such as total quality management (TQM), best practice benchmarking (BPB), e-government through business process reengineering (BPR), performance management (PM), and others where technology is the primary driver for the reforms and optimism for better *competing* results. The focus on "how-to" solutions has boldly undermined the public administrator's broader *public service* role in a democratic society.

The renewed sense of urgency is for public administrators to ask the "*why*" question: Why do certain programs fail and do not produce what is expected in others? Why do certain actions not produce better results and yet swarms of clients are ready to follow? To answer the why question is to interpret the information learned from practices. It is to connect and reconnect the dots by imaginizing the real-life stories as they unfold from formal data. We must understand that data is not a "thing" or a "statistic" that is manipulable—up or down, pass or fail, hit or miss, dead or alive—they are formative events or results that have a history. We must be interested in that history to be informed about values that bind us, even though they may be in the remotest part of the world or even in the most undemocratic of places where women and children's rights are held captive under formal cultural garb. The informality should give us a clue to the formal behavior, just as informal practice is evidence of the soundness of a theory. Rather than asking the question *how* information technology can be used here to answer the *why* question, we should turn the question on its head: "*Why* cannot information technology be used to answer such questions?" Public administrators should be taking advantage of the emerging technologies to be informed about informal practices in order to contextualize and learn from them. They should also be encouraged to participate in developing better *lenses* (tools) that can show us the pathways to interpret other's practices.

Accordingly, the challenge against bureaucratic norms is not against public service, but against the organized action of public agencies that constrain democratic practice.[4] The idea that governments must gather information from citizens in order to serve them better is a collective myth because, as

Practice

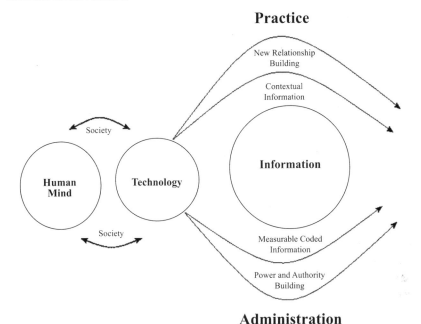

Figure 6. Practice-administration dichotomy in the Information Age

argued here, pertinent information cannot be parsed and then pieced to-gether to create a product called service. The current paradigm of informa-tion gathering in public administration includes the presumption that the information will be used as an ingredient (input) to create a product called service. Service is a call, not a thing that can be produced without putting that into practice. In order to be empowered, public administrators do not strictly need data and information; they need the ability to translate the in-formation into action.

The implication of the practice-administration dichotomy is explained using a schematic logic as in figure 6. Human rationality is not independent of the social order; it is shaped by social interaction. Following the work of Michael Gazzaniga (2011b), I argue that social responsibility arises out of so-cial interaction and that the mind constrains the brain. Given that our mental life is codependent upon, and therefore constrained by, the social order, we tend to liberate ourselves by using technology to communicate and translate our information to influence or motivate social behavior. Technology is not independent of one's worldview; however, that is not the primary focus here. The *type* of information and the *amount* of information that is transmitted using a given technology is critical to the message that is being communicated to others. In other words, not all information is the same: *how I say it and how much I say matters.* I elaborate these points as follows:

First, I divide information into measurable information (for example, standardized data) and contextual information (for example, situation-specific data such as in narratives, videos, pictures, and such). Measurable information is necessary to organize, administer, and monitor social behavior. In this regard, technology is applied to keep the democratic system operating routinely. What I call *administration*, when it comes to information, is to routinize information to minimize the unpredictability and maintain or standardize the information among its constituent group at a constant pace. On the other hand, contextual information is necessary for putting information into *practice*. Experiential learning has to be contextualized in order to enable meaningful action.

Second, technology as the medium for communication can enable the flow of information to continue or can cut the flow, depending on how much information is being transmitted. This depends on whether the technology is being used to *distribute power* or *maintain existing power relations*. Technology is an enabler of distributive power as it builds on relational networks. It can also be a constrainer of power as it controls and maintains the power relations.

It might be instructive to explain the relationship between information and power. Garfinkel (2008) defines power as "the probability that communicant A can by his communicative work so alter communicant B's definition of his situation as to restrict B's alternatives of conduct to those that A desires" (Garfinkel, 2008, p. 225). If information is used to restrict or even alter someone's own desires and preferences in favor of the communicant's (the informer's) desires and preferences so as to limit the receiver's ability to change the situation on their own, then power has been used, essentially, for one to dominate over the other. Therefore, the amount of information is restricted and reduced to an order or instruction to maintain control and authority for the greater good. The constitutive feature of communication is restricted in order to rationalize the commands for administering and controlling unexpected and undesirable behaviors. Information in this regard is used narrowly for the sake of control. In contrast, however, when information is provided with the intent to distribute authority and build relationships, the constitutive feature of communication is maintained and the information receiver is not deprived of the information that carries values. The receiver is not subjected to *information rationing* by the informant; rather, the immutable information is preserved and passed on to the receiver for the greater good. Therefore, the decision (or order) does not precede the action;

rather, the action/practice is part of the information provided to make decisions. Harmon and White (1989) clearly captured the idea when they stated: "Rather than thought preceding action (linked by decisions), thought and action are mutually constitutive and coextensive. Decisions are not objectively real but are objectifications of the ongoing flow of social process. Informally, decisions may be thought of as 'stopped processes'" (p. 146).

Practice-based information (what I also identify as contextual information) incorporates the *unexpected* as well as the *expected* into the decision model. Both are essentially the constitutive part of that order. Rather than controlling for what information could be passed on (with the intention to minimize undesirable and unpredictable outcomes), the practice-based information, even at the most informal level, becomes a resource for the individual receiving the information that could then be used for meeting one's preferences and desires without pressure or prejudice. The value of contextual information is in having faith that individuals can make their own decisions based on experiential learning from others. For Garfinkel, as Ann Rawls notes, "information is to be found in the presence of anomaly" (Garfinkel, 2008, p. 20). To learn from practice, we have to expect what is *unexpected* and be open to learning from what was yet *unknown*. Pretending to have knowledge is more dangerous than ignorance because learning stops for the perceived omniscient. The underlying precept for practice-based information is the building of trust and commitment between the informer and the informed based on their commitment to a particular, local, situated order. When asked how to build trust, James March responded, "Trust is not a calculated exchange that I trust you because you trusted me, rather for trust to be anything truly meaningful, you have to *trust somebody who isn't trustworthy*. Otherwise, it's just a standard rational transaction. The relationships among leaders and those between leaders and their followers certainly involve elements of simple exchange and reciprocity, but humans are capable of, and often exhibit, more arbitrary sentiments of commitment to one another" (Coutu, 2006, p. 86; emphasis added).

The practice-administration dichotomy explains the tension between maintaining the status quo and creating what is new and unknown. How we rationalize norms to control and maintain the status quo through measurable objectives is one issue, and how we contextualize information and empower to create what is yet unknown is another issue. The emphasis here is not on choosing practice over administration or vice versa. In a democracy, the balance has to be between maintaining democratic rights (administration) and enabling a culture where democratic values can be exemplified by action (practice). We ought to preserve one without undermining the value of the other. The tension is real, and maintaining a balance is profound. This should remind us of George Washington's dictum in his 1796 farewell address: "Your union ought to be considered as a main prop of your liberty, and that the love of the one ought to endear to you the preservation of the other" (Washington, 2008/1796).

The implication of the practice-administration dichotomy is of great significance in the Information Age of the 21st century. As discussed in this book, there are obvious costs to individuating information and rationalizing the usage of it based on a prescribed social order. Objectifying the goals through measurable outcomes may well serve the purpose of maintaining as well as controlling a stable and functioning democracy but not without the cost of losing trust in government. Technology has only made it easier to collect and monitor the information necessary for the functions of a stable democracy. The dominance of technology for developing techniques and the sheer dominance of Internet-based data gathering should be of concern for any citizens who care for democratic ideals. However, the upside of the Information Age is that technology has also opened unprecedented opportunities for social mobilization by connecting individuals through social contact. We are at a peak in the Internet generation where information becomes a viable asset for learning from practice. Where information without context can be a liability and responsibility (because of faulty interpretation), information with context is a resource for learning.

The more we invest in the administrative side of governance to maintain the status quo, the more we alienate citizens and lose their trust. In contrast, the more we pay attention to informal practice-based governance, the more we bring citizens closer to the government. The one who wishes to incorporate practice into administration becomes a walking theoretician. For public administrators, the idea would be to use information technology for building social relations so that we can learn from practice. Therefore, one of the fundamental uses of information technology must be geared toward increasing the social value of information. In other words, information should be used to form interorganizational networks that increase participants' trust.[5] We must take full advantage of the emerging social networking tools such as Facebook, Google Talk, Skype, Twitter, blogs, Wiki, YouTube, Vimeo, and other mobile-based tools including phones—now everyday tools that help us learn from other's practices and values so we can *take charge of emerging situations*. Since they are relational tools, they have the potential to inform us through contextualization about the values that bind us to each other within societies.

Indeed, democratic values emerge from practice. Practice-based learning is value-centric; values are embedded within practices. To administer in a democracy is first to understand emerging values, and then act on those values (Frederickson, 1997). It is not as much about the *destination* as it is about the *journey*. As public administrators devote more of their energy to administration than to practice, they lose the connection to their constituents, making government appear nonresponsive and often irresponsible to citizen aspirations. Ultimately, the price is paid as citizens start losing trust in their government, and government makes irresponsible choices that further alienate citizens and governments. Despite ever-improving technologies in the Information Age, the alienation of citizens and government will continue unless we steer technology toward learning values from practice—the Art of Administration, and while doing so conserve rights and liberties through the Science of Administration. The emerging values are embedded in practices, and managers must be on top of it. "Democracy will not be delivered, renewed or upgraded automatically, like the latest Netflix blockbusters through our broadband connections and smart phones. The future of freedom in the Internet age depends on whether people can be bothered to take responsibility for the future and act. Just as our individual actions and choices as citizens, parents, teachers, employees, managers, and government officials combine to shape the kind of world we live in, the actions and choices of each and every one of us are shaping the internet's future" (MacKinnon, 2012, p. 222).

The central question in the modern world should not be about technology and its effectiveness in building better democracies; rather, the question that should be asked is if the technologies we use are effective in helping understand shared values and practices to tackle the world's most pressing needs. When the government is not engaged in nurturing social values, it exposes the limits of democracy. For example, modern democracies cannot hide but must address the fundamental issues of social justice and increasing disparity in income and the rising social underclass. We have a choice: We can either use technological means to understand social values and pave the way for social empowerment, or we can use technological means to find the cause and inject the solution. While the latter approach will surely fulfill the *objective* (or ego) of the policy maker, the former approach will likely earn the trust of a partner who can then help in *finding* a solution. Indeed, scientific rationalism in the affairs of government can be a strong influence in turning democracies into demagogueries.

In recent years, social entrepreneurs have taken the lead that regenerated interest in government and civil society organization partnerships to solve the world's most pressing problems, including establishing democratic rights of individuals, coping with climate change, rights to healthy living, and social justice for marginalized communities (see Haque, 2013).[6] What is instructive from the successes of civil society enterprises in the developing world is that trust building is the precondition to learning. It must be noted that the successes of social entrepreneurs in alleviating poverty and removing social injustices have achieved more through building social relations via mobile technologies and rudimentary face-to-face social interactions, and less through monitoring and controlling activity. Social entrepreneurs used technology for communities to meet, connect, and build cooperative enterprises through human relations. Noble laureate Elinor Ostrom may have captured this early on when she noted, "face-to-face communication in a public good game—as well as in other types of social dilemmas—produces substantial increases in cooperation" (Ostrom, 2000, p. 140). She further exclaimed, "When communication is implemented by allowing subjects to signal promises to cooperate through their computer terminals, much less cooperation occurs than in experiments allowing face-to-face communication" (Ostrom, 2000, pp. 140–141).[7] If the tool is perceived as the means to an end then cooperation is unlikely to occur. This is because, as Garfinkel reminded us, when communication is symbolic, nonrelational, it is preprogrammed, rather than being free. Just as in language, information has to become a self-regulating order of constitutive practice.

Steve Jobs, founder of Apple Inc., once said, "Technology alone is not enough. It is technology married with liberal arts, married with humanities, that yields the results that make our hearts sing" (Economist, 2011). Technology by its very nature does not have a purpose to serve unless it somehow reflects human values. According to Krista Tippett (2010), "Albert Einstein, the father of modern science, in his later years lamented that 'science in his generation has become a razor blade in the hands of a three year old child,' and he stressed the importance of virtues more, not less, in a technologically advanced society. He opined, 'These kinds of people are geniuses in the art of living more necessary to the dignity, security and the joy of humanity than the discoveries of objective knowledge'" (Tippett, 2010).

Technology becomes powerful when it becomes commonplace. Information technology is now inseparable from democratic institutions. We must take full advantage of technology to defend and strengthen democratic institutions upon which freedom and progress depend. Where we invest technology and what kind of technology we implement in government should be matters of concern for citizens and governments alike. Just as electronic government adds to the convenience of delivering public services to our doorsteps, it can also be used to strengthen political regimes (Ahn & Bretshneider, 2011) or solidify corrupt practices within weak political institutions (Lindstedt & Naurin, 2010). I have explained in this work that if we decide to invest technology skills and resources towards *administration*, we will increasingly alienate citizens from government. On the other hand, if we invest technology in learning from *practice*, we can bring citizens closer to government. This practice-administration dichotomy presents us with a dilemma that we should present to every public administrator dealing with technology in the Information Age. Whether our regime's activities should exemplify control-oriented values or development-oriented values[8] should be a valid question for the millennium generation. The weight of either set of values must be balanced as it will fundamentally influence the emerging political culture. The imbalance in values will ultimately shift our priorities, forcing us to move into territories where emerging values are not compatible with democratic values.

Notes

Chapter 2

1. The instrumental view of public administration has recently received wide criticism within public administration (see Spicer, 1995, 2007; Haque & Spicer, 1997; Haque, 1997).

2. Edmund Burke's worldview has received wide attention in public administration because of the implications of his ideas for American public administration rooted in constitutional principles (see Rohr, 1986; Terry, 1995; Haque & Spicer, 1997, Haque, 1998, 2004).

3. Figure 1 found in the introduction should be instructive here. I discuss where technique can be most appropriate given the institutional goals and the skill requirements to adapt to technology.

4. What is more profound, according to Marcuse, is our constant mediation between nature (tradition, society, natural world) and liberation. Whereas nature is rooted in ideology (deep rooted tradition), liberation is rooted in science and reason. Our longing for liberation through technological power destroys the very nature we want to preserve. At the end, technology wins in perpetuity to liberate the larger society at the expense of the destruction of nature. Although nature loses to technological supremacy, the elite in the political machinery ensure that a part of the original nature (not destroyed) is still available for themselves (and away from the general citizens). Therefore, in the process of human mediation between the technological and the natural world, one privileged class gets the best of both—nature and liberation. Marcuse (1964) noted, "The technological transformation is at the same time political transformation, but the political change would turn into qualitative social change only to the degree to which it would alter the direction of the technological progress—that is, develop a new technology. For the established technology has become an instrument of destructive politics" (p. 227).

5. The financial system that is set up through the grand EA plan essentially

connects all federal agencies' independent financial management systems (FMS). A list of 23 FMS is found in the report (see CFO Report to Congress, Office of Management and Budget, 2011).

6. For the implications of Garfinkel's ideas for public administration, see Harmon and White (1989).

7. In an organizational setting, Feldman and Pentland (2003) refer to this as the "ostensive" part of the routine. A good discussion about operationalizing ostensive and performative can be found in Strum and Latour (1999). They discuss how baboons as social players actively negotiate and renegotiate "to make their society" (in contrast to our traditional ostensive definitions gathered from data).

Chapter 3

1. Anthony Giddens (1986) uses the words "'go on' within the routines of social life" (p. 4).

2. A good discussion about intermediaries and mediator can be found in Latour (2005), pp. 37–42.

3. Latour's quote relates to his discussion about the "discovery" and mapping of Sakhalin Island by the French navigator Laperouse in 1787 through Chinese fishermen. Laperouse wanted to know whether Sakhalin was an island or a peninsula. He encountered the local people who drew him a map on sand, which was then copied on paper and transported back to Europe as proof that Sakhalin was indeed an island according to the local people's understanding and knowledge of the area. Ten years later in 1797 the English ship *Neptune* went on the same expedition, but this time they were *stronger* given they had the maps (inscribed from Laperouse's crude map ten years before) that they didn't have before, including log books, nautical instructions, and such. Knowledge had been accumulated and mobilized to make it appear the Europeans were now stronger (or smarter). For more details, see Latour (1987), chapter 6.

Chapter 4

1. A comprehensive list of global poverty statistics can be found in the UN World Food Program (2015).

2. For large-scale IT projects in the private sector the failure rate is between 35 and 75%. See McAfee (2003).

3. A good example of situational order can be found in the emergency management practices of Mayor Walt Maddox of Tuscaloosa during the April 2011 tornado. See the *New York Times* article by Severson and Brown (2011).

4. In the business world this phenomenon is common. For example, think about having multiple credit cards and having your name in multiple databases and then wonder why you receive more telemarketing calls than someone who

has no credit cards. Or, you give donations online and start receiving similar donation requests from multiple nonprofit organizations.

5. Recently in China, a name chosen by parents for their newborn was rejected because the name did not exist in the official database of names of Chinese residents.

Chapter 5

1. Garfinkel (2008) explains the noises as the "randomness or meaningless of a set of signals—the unidealizeable [sic] character of a set of signals. Noise refers to the extent to which a communicant can invoke a set of constant possibilities as standards for effecting the idealization of signals" (p. 183).

2. For example, democracy has limited meaning until individuals in a democractic society practice it. There is no universal meaning of democracy; it is a word that defines the concept. But to realize the concept we have to live in that democracy to know the value of democracy and how it is realized by the people who claim to be the citizens of that democracy.

3. How technology as an artifact can permeate social norms and is used to reinforce social order can be understood with a narrative from a personal recollection of a story I heard from a colleague several years ago: Living in a neighborhood close to the railroad tracks, the young boy was asked by his father never to cross to the other side of the tracks because it was a dangerous place ruled by strangers. After several years the young boy, now an adult, decided to placate his curiosity by crossing the railroad tracks to meet the strangers. What he found not only shocked him but also shocked the people he met. The strangers happened to be black mine workers and their families living in the neighborhoods across the tracks in the late 1940s in Birmingham, Alabama. The perception of the young white boy about the dangerous place was implanted with the aid of an artificial demarcation created by technology—the railroad tracks. Although the railroads built in the early 1900s were never meant to segregate neighborhoods, the tracks became a convenient tool for expressing the frame of mind that dominated the South at that time, and they reinforced the social demarcation of neighborhoods based on race and social class. The startling contrast of the human condition between the two sides of the railroad tracks today, for hundreds of miles across the South, is a riveting reminder that what humans cannot do in a thousand years, technology can effect in less than half a century. Whereas a young boy found the strangers to be ordinary, normal people going about their lives, the black mine workers were filled with curiosity and fear about what brought the young white soul to their side of the tracks while hoping, perhaps he is one of them! Indeed, in another place, time, or circumstance, they perhaps would all have lived side by side or even in the same house. What is perceived to be an object—the railroad track—meant different things to dif-

ferent people. To the young boy, it meant the barrier of the danger zone; to a black mine worker it was the line that set him apart from the rest of the prosperous world; and to the young boy's dad, it was a convenient tool to keep his family away from the black people. Despite the oddity of the situation, life was stable on both sides of the railroad tracks because all were committed to their "self-regulated practices." An artifact of technology was just one among other means to express the social order.

4. The age of bureaucracy is also the "era of the information society," claimed Christopher Dandeker (1990, p. 2). In order to organize, we have to observe, gather, and systematically monitor individuals. Routine surveillance is a prerequisite of effective social organization (Giddens, 1987) that has put democracy in a state of surveillance in the information society.

5. This is clearly echoed by Jane Fountain (2001), one of the pioneers who wrote about the impact of the Internet on bureaucracy (see pp. 78–82).

6. In the Grameen Bank project by Nobel laureate Dr. Mohammad Yunus, mobile technologies have been successfully applied to community empowerment, social emancipation, and social justice by providing marginalized communities, particularly women, with access to markets, government services, and avenues to participate in the democratic process (Yunus & Weber, 2007, 2010; Martin & Osberg, 2007; Bhatnagar, 2000).

7. Recent technological breakthroughs make it possible for large participants to gather face to face to deliberate in virtual settings. Amitai Etzioni (2004), citing several national surveys, expressed his optimism that online communities can make people more social than offline communities, as people have far more opportunities to interact while conducting business or connecting to friends and family online. He notes that we may not be able to create new relationships due to trust issues; however, online communities can reinforce offline ones. He made the arguments before Facebook made its mark and before the Middle East revolutions ("Arab Spring") that were triggered by online social mobilization of the masses.

8. According to Rokeach (1973), control-oriented values include discipline, efficiency, responsibility, punctuality, and so on. Development-oriented values include constitutional principles, values of freedom, creativity, integrity, and trust. See also Haque (2005, p. 484).

References

Adams, G. B., & Balfour, D. L. (1998). *Unmasking Administrative Evil*. Thousand Oaks, CA: Sage.

Adams, G. B., & Balfour, D. L. (2009). Ethical failings, incompetence, and administrative evil: Lessons from Katrina and Iraq. In R. W. Cox (Ed.), *Ethics and Integrity in Public Administration: Concepts and Cases* (pp. 40–64). Armonk, NY: M. E. Sharpe.

Ahmad, P. (1999). Aesthetics and vocabulary of Nakshi Kantha. *Vihangama: The IGNCA Newsletter, VII* (1–4). Retrieved from http://ignca.nic.in/nl_body.htm

Ahn, M. J., & Bretschneider, S. (2011). Politics of e-government: E-government and the political control of bureaucracy. *Public Administration Review, 71* (3), 414–424. doi:10.1111/j.1540–6210.2011.02225.x

Alegria, M. (2009). Training for research in mental health and HIV/AIDS among racial and ethnic minority populations: Meeting the needs of new investigators. *American Journal of Public Health, 99 (Suppl. 1),* S26–30. doi:10.2105/ajph.2008.135996

Alexander, J. (1992). The promise of a cultural sociology: Technological discourse and the sacred and profane information machine. In R. Münch & N. J. Smelser (Eds.), *Theory of Culture* (pp. 293–323). Berkeley, CA: University of California Press.

Alkadry, M. G., & Blessett, B. (2010). Aloofness or dirty hands? Administrative culpability in the making of the second ghetto. *Administrative Theory & Praxis, 32* (4), 532–556. doi:10.2753/ATP1084–180632403

Archibald, J. (2012, February 15). Hoover don't know Squanto: Police marking people of Asian Indian descent as Native American. *Birmingham News*. Retrieved from http://blog.al.com/archiblog/2012/02/hoover_dont_know_squanto.html

Asch, S. E. (1951). Effects of group pressure upon the modification and distortion of judgments. In H. S. Guetzkow (Ed.), *Groups, Leadership, and Men* (pp. 177–190). Pittsburgh, PA: Carnegie Press.

Ayres, I. (2007). *Super Crunchers: Why Thinking-by-Numbers Is the New Way to Be Smart*. New York, NY: Bantam Books.

Barnard, C. I. (1968). *The Functions of the Executive*. Cambridge, MA: Harvard University Press.

Behn, R. (1996). Public management: Should it strive to be art, science, or engineering? *Journal of Public Administration Research and Theory, 6* (1), 91–123.

Bellah, R. N. (2008). *Habits of the Heart: Individualism and Commitment in American Life*. Berkeley, CA: University of California Press.

Berners-Lee, T. (1989, March). *Original proposal for a global hypertext project at CERN*. Retrieved February 25, 2014 from http://www.w3.org/History/1989/proposal.html

Bhatnagar, S. C. (2000). Social implications of information and communication technologies in developing countries: Lessons from Asian success stories. *Electronic Journal of Information Systems in Developing Countries, 1* (4), 1–9. Retrieved from http://www.ejisdc.org

Boer, T., Pastor, M., Sadd, J. L., & Snyder, L. D. (1997). Is there environmental racism? The demographics of hazardous waste in Los Angeles County. *Social Science Quarterly, 78* (4), 793–810.

Boudreau, M.-C., & Robey, D. (2005). Enacting integrated information technology: A human agency perspective. *Organization Science, 16* (1), 3–18. doi:10.1287/orsc.1040.0103

Bourdieu, P. (1989). Social space and symbolic power. *Sociological Theory, 7* (1), 14–25.

Bourdieu, P. (1990). *The Logic of Practice*. Stanford, CA: Stanford University Press.

Bovens, M., & Zouridis, S. (2002). From street-level to system-level bureaucracies: How information and communication technology is transforming administrative discretion and constitutional control. *Public Administration Review, 62* (2), 174–184. doi:10.1111/0033-3352.00168

Bowker, G., & Star, S. (1999). *Sorting things out: Classification and its consequences*. Cambridge, MA: MIT Press.

Bozeman, B., & Bretschneider, S. (1986). Public management information systems: Theory and prescription. *Public Administration Review, 46* (special issue), 475–487.

Bretschneider, S., & Wittmer, D. (1993). Organizational adoption of microcomputer technology: The role of sector. *Information Systems Research, 4* (1), 88–108. doi:10.1287/isre.4.1.88

Brown, M. M. (2003). Technology diffusion and the knowledge barrier: The dilemma of stakeholder participation. *Public Performance & Management Review, 26* (4), 345–359.

Brown, M. M., & Brudney, J. L. (2003). Learning organizations in the public sector? A study of police agencies employing information and technology to advance knowledge. *Public Administration Review, 63* (1), 30–43. doi:10.1111/1540-6210.00262

Bruner, J. (2011). Where Americans are moving (interactive map). Retrieved October 2011 from http://www.forbes.com/2010/06/04/migration-moving -wealthy-interactive-counties-map.html

Bryer, T. A. (2011). The costs of democratization: Social media adaptation challenges within government agencies. *Administrative Theory & Praxis, 33* (3), 341–361. doi:10.2753/ATP1084-1806330302

Bryer, T. A., & Zavattaro, S. M. (2011). Social media and public administration: Theoretical dimensions and introduction to the symposium. *Administrative Theory & Praxis, 33* (3), 325–340. doi:10.2753/ATP1084-1806330301

Callon, M. (1986). Some elements of a sociology of translation: Domestication of the scallops and the fishermen of St Brieuc Bay. In J. Law (Ed.), *Power, Action, and Belief: A New Sociology of Knowledge* (pp. 196–233). London, England: Routledge & Kegan Paul.

Catalani, C., & Minkler, M. (2010). Photovoice: A review of the literature in health and public health. *Health Education & Behavior, 37* (3), 424–451. doi:10.1177/1090198109342084

Chandler, R., & Adams, B. (1997). Let's go to the movies! Using film to illustrate basic concepts in public administration. *Public Voices, 3* (1), 9–26.

Clair, R. P. (2003). *Expressions of Ethnography: Novel Approaches to Qualitative Methods.* Albany, NY: State University of New York Press.

Cohen, P. (2009, October 19). Field study: Just how relevant is political science? *New York Times.* Retrieved from http://www.nytimes.com

Coleman, S. (2005). Blogs and the new politics of listening. *Political Quarterly, 76* (2), 272–280. doi:10.1111/j.1467-923X.2005.00679.x

Cooper, T. L. (1990). *The Responsible Administrator: An Approach to Ethics for the Administrative Role.* San Francisco: Jossey-Bass Publishers.

Cooper, T. L. (2011). Citizen-driven administration: Civic engagement in the United States. In D. C. Menzel & H. L. White (Eds.), *The State of Public Administration: Issues, Challenges, and Opportunities* (pp. 238–256). Armonk, NY: M. E. Sharp.

Coutu, D. (2006). Ideas as art: A conversation with James G. March. [Interview]. *Harvard Business Review, 84* (10), 82–89.

Dandeker, C. (1990). *Surveillance, Power, and Modernity: Bureaucracy and Discipline from 1700 to the Present Day.* New York, NY: St. Martin's Press.

Davidson, C. N. (2011). *Now You See It: How the Brain Science of Attention Will Transform the Way We Live, Work, and Learn.* New York, NY: Viking.

Denhardt, R. B., & Denhardt, J. V. (2000). The new public service: Serving rather than steering. *Public Administration Review, 60* (6), 549–559. doi: 10.2307/977437

Denrell, J., & March, J. G. (2001). Adaptation as information restriction: The hot stove effect. *Organization Science, 12* (5), 523–538.

Dodge, J., Ospina, S. M., & Foldy, E. G. (2005). Integrating rigor and relevance in public administration scholarship: The contribution of narrative inquiry. *Public Administration Review, 65* (3), 286–300. doi:10.1111/j.1540-6210.2005.00454

Drengson, A. R. (1981). The virtue of socratic ignorance. *American Philosophical Quarterly, 18* (3), 237–242. doi: 10.2307/20013918

Economist. (2010, February 27). Data, data everywhere [special section]. *The Economist, 394* (8671), 3–5.

Economist. (2011, October 6). Steve Jobs. *The Economist* (online). Retrieved from http://www.economist.com/blogs/babbage/2011/10/obituary

Ellul, J. (1964). *The Technological Society.* New York, NY: Knopf.

Elwood, S. (2010). Geographic information science: Emerging research on the societal implications of the geospatial web. *Progress in Human Geography, 34* (3), 349–357.

Elwood, S. (2011). Geographic Information Science: Visualization, visual methods, and the geoweb. *Progress in Human Geography, 35* (3), 401–408. doi: 10.1177/0309132510374250

Ertel, W. (2011). *Introduction to Artificial Intelligence.* New York, NY: Springer.

ESRI. (2011a). *GIS Stimulus Reporting: GIS for State Governments.* Retrieved October 2011 from http://www.esri.com/industries/stategov/business/stimulus.html

ESRI. (2011b). *Mapping for Everyone: Express Yourself with Maps.* Retrieved January 2012 from http://www.esri.com/mapping-for-everyone/

Etzioni, A. (2004). On virtual, democratic communities. In A. Feenberg & D. D. Barney (Eds.), *Community in the Digital Age: Philosophy and Practice* (pp. 225–238). Lanham, MD: Rowman & Littlefield.

Feenberg, A. (1991). *Critical Theory of Technology.* New York, NY: Oxford University Press.

Feldman, M. S. (2004). Resources in emerging structures and processes of change. *Organization Science, 15* (3), 295–309. doi:10.1287/orsc.1040.0073

Feldman, M. S., & March, J. G. (1981). Information in organizations as signal and symbol. *Administrative Science Quarterly, 26* (2), 171–186.

Feldman, M. S., & Pentland, B. T. (2003). Re-conceptualizing organizational routines as a source of flexibility and change. *Administrative Science Quarterly, 48* (1), 94–118.

Feldman, M. S., Skoldberg, K., Brown, R. N., & Horner, D. (2004). Making sense of stories: A rhetorical approach to narrative analysis. *Journal of Public Administration Research and Theory, 14* (2), 147–170. doi:10.1093/jopart/muh010

Floridi, L. (2010). *Information: A Very Short Introduction.* Oxford; New York: Oxford University Press.

Follett, M. P. (1942). *Dynamic Administration: The Collected Papers of Mary Parker Follett.* Henry C. Metcalf and L. Urwick (Eds.). New York, NY: Harper & Row.

Forbes Magazine (2011). American migration [interactive map]. *Forbes,* n.d. Web. 2011. Retreived from http://www.forbes.com/special-report/2011/migration.html

Foucault, M. (1977). *Discipline and Punish: The Birth of the Prison*. New York, NY: Pantheon Books.

Fountain, J. E. (2001). *Building the Virtual State: Information Technology and Institutional Change*. Washington, DC: Brookings Institution Press.

Fox, S., & Rainie, L. (2014, February 27). *The Web at 25 in the U.S.* Retrieved from Pew Research Center website: http://www.pewinternet.org/files/2014/02/PIP_25th-anniversary-of-the-Web_022714_pdf.pdf

Frederickson, H. G. (1980). *New Public Administration*. Tuscaloosa, AL: University of Alabama Press.

Fredrickson, H. G. (1990). Public administration and social equity. *Public Administration Review, 50* (2), 238–237.

Frederickson, H. G. (1997). *The Spirit of Public Administration*. San Francisco, CA: Jossey-Bass.

Fung, A., Graham, M., & Weil, D. (2007). *Full Disclosure: The Perils and Promise of Transparency*. New York, NY: Cambridge University Press.

Gadamer, H.-G. (1975). *Truth and Method*. New York, NY: Seabury Press.

Gallagher, S. (1998, March 30). Beat the systems management odds. *Information Week, 675*, 61–76.

Garfinkel, H. (1967). *Studies in Ethnomethodology*. Englewood Cliffs, NJ: Prentice-Hall.

Garfinkel, H. (2008). Toward a sociological theory of information. A. W. Rawls (Ed.), *Toward a Sociological Theory of Information* (pp. 101–304). Boulder, CO: Paradigm.

Gartner Group. (2011, August 4). Gartner says worldwide mobile connections will reach 5.6 billion in 2011 as mobile data services revenue totals $314.7 billion. Retrieved from Gartner website: http://www.gartner.com/it/page.jsp?id=1759714

Gartner Group. (2012). Data mining. In *IT Glossary*. Retrieved March 2012 from http://www.gartner.com/it-glossary/data-mining/

Gazzaniga, M. S. (2011a). Who is in charge? *BioScience, 61* (12), 937–938. doi: 10.1525/bio.2011.61.12.2

Gazzaniga, M. S. (2011b). *Who's In Charge? Free Will and the Science of the Brain*. New York, NY: HarperCollins.

General Service Administration. (2011, October 7). *Narrative Networks* (DARPA-BAA-12-03). Retrieved from FedBizOpps.gov website: https://www.fbo.gov/index?s=opportunity&mode=form&id=fd625a4022ec38fde2a8f6f1f4628395&tab=core&_cview=0

Giddens, A. (1979). *Central Problems in Social Theory: Action, Structure, and Contradiction in Social Analysis*. Berkeley, CA: University of California Press.

Giddens, A. (1984). *The Constitution of Society: Outline of the Theory of Structuration*. Berkeley, CA: University of California Press.

Giddens, A. (1986). *The Constitution of Society: Outline of the Theory of Structuration*. Berkeley: University of California Press.

Giddens, A. (1987). *A Contemporary Critique of Historical Materialism: Vol. 2 The Nation-State and Violence*. Berkeley, CA: University of California Press.

Giddens, A. (1993). *New Rules of Sociological Method: A Positive Critique of Interpretative Sociologies* (2nd ed.). Stanford, CA: Stanford University Press.

Goguen, J. (1992). *The Dry and the Wet*. Oxford, England: Oxford University Computing Laboratory, Programming Research Group.

Goguen, J. (1997). Towards a social, ethical theory of information. In G. C. Bowker, S. L. Star, & L. Gasser (Eds.), *Social Science, Technical Systems, and Cooperative Work: Beyond the Great Divide* (pp. 27–56). Mahwah, NJ: Lawrence Erlbaum Associates.

Goldfinch, S. (2007). Pessimism, computer failure, and information systems development in the public sector. *Public Administration Review, 67* (5), 917–929. doi:10.1111/l.1540–6210.2007.0078.x

Goodsell, C. T. (1992). The public administrator as artisan. *Public Administration Review, 52* (3), 246–253.

Goodsell, C. T., & Murray, N. (1995). Prologue: Building new bridges. In C. T. Goodsell & N. Murray (Eds.), *Public Administration Illuminated and Inspired by the Arts* (pp. 3–23). Westport, CT: Praeger.

Government Accountability Office. (2011). *Data Mining: DHS Needs to Improve Oversight of Systems Supporting Counterterrorism* (GAO 11–742). Retrieved from http://www.gao.gov/products/GAO-11-742

Gross, G. (2013). Healthcare.gov had no chance in hell. *Computerworld, 47* (19), 6–6.

Hallisey, E. J. (2005). Cartographic visualization: An assessment and epistemological review. *Professional Geographer, 57* (3), 350–364. doi:10.1111/j.0033-0124.2005.00483.x

Haque, A. (1997). Human nature, tradition and law: A Burkean perspective in public administration. *Journal of Management History, 3* (3), 256–271.

Haque, A. (2001). GIS, public service, and the issue of democratic governance. *Public Administration Review, 61* (3), 259–265. doi:10.1111/0033-3352.00028

Haque, A. (2003). Information technology, GIS, and democratic values: Ethical implications for IT professionals in public service. *Ethics and Information Technology, 5* (1), 39–48. doi:10.1023/A:1024986003350

Haque, A. (2004). The concept of unified administration in a democratic republic. *Administration & Society, 35* (6), 701–716. doi:10.1177/0095399703256775

Haque, A. (2005). Information technology and surveillance: Implications for public administration in a new world order. *Social Science Computer Review, 23* (4), 480–485. doi:10.1177/0894439305278874

Haque, A., & Mantode, K. (2013). Governance in the technology era: Implications of actor network theory for social empowerment in South Asia. In Y. Dwivedi, H. Henriksen, D. Wastell & R. De' (Eds.), *IFIP Advances in Information and Communication Technology: Vol. 402. Grand Successes and Failures in IT: Public and Private Sectors* (pp. 375–390). New York, NY: Springer. doi:10.1007/978-3-642-38862-0

Haque, A., & Spicer, M. W. (1997). Reason, discretion, and tradition: A reflection on the Burkean worldview and its implications for public administration. *Administration & Society, 29* (1), 78–96. doi:10.1177/009539979702900105

Harmon, M. M. (2006). *Public Administration's Final Exam: A Pragmatist Restructuring of the Profession and the Discipline.* Tuscaloosa, AL: University of Alabama Press.

Harmon, M. M., & McSwite, O. C. (2011). *Whenever Two or More Are Gathered: Relationship as the Heart of Ethical Discourse.* Tuscaloosa, AL: University of Alabama Press.

Harmon, M. M., & White, J. D. (1989). "Decision" and "action" as contrasting perspectives in organization theory. *Public Administration Review, 49* (2), 144–152.

Hayek, F. A. (1983). *Knowledge, evolution, and society.* London: ASI (Research).

Heaton, B. (2011, October 21). Facial recognition technology spurs privacy concerns for Feds. *Government Technology Magazine.* Retrieved from http://www.govtech.com

Heeks, R., & Stanforth, C. (2007). Understanding e-Government project trajectories from an actor-network perspective. *European Journal of Information Systems, 16* (2), 165–177. doi: 10.1057/palgrave.ejis.3000676.

Heidegger, M. (1977). *The Question Concerning Technology, and Other Essays.* New York, NY: Harper & Row.

Hennessy, E., Kraak, V. I., Hyatt, R. R., Bloom, J., Fenton, M., Wagoner, C., & Economos, C. D. (2010). Active living for rural children: Community perspectives using PhotoVOICE. *American Journal of Preventive Medicine, 39* (6), 537–545. doi:10.1016/j.amepre.2010.09.013

Hummel, R. P. (1990). Uncovering validity criteria for stories managers hear and tell. *American Review of Public Administration, 20* (4), 303–314.

Hummel, R. P. (1991). Stories managers tell: Why they are as valid as science. *Public Administration Review, 51* (1), 31–41.

Ihde, D. (1990). *Technology and the Lifeworld: From Garden to Earth.* Bloomington, IN: Indiana University Press.

Internet World Stats (2014). World Internet users statistics and 2014 world population stats. Miniwatts Marketing Group, n.d. Web. Accessed January 8, 2015. Retrieved from http://www.internetworldstats.com/stats.htm

Janis, I. L. (1972). *Victims of Groupthink; A Psychological Study of Foreign-Policy Decisions and Fiascoes.* Boston, MA: Houghton Mifflin.

Jasīmauddīn & Milford, E. M. (1939/1958). *The Field of the Embroidered Quilt: A Tale of Two Indian Villages.* Calcutta, India: Oxford University Press.

Kaghan, W. N., & Bowker, G. C. (2001). Out of machine age: Complexity, sociotechnical systems, and actor network theory. *Journal of Engineering and Technology Management, 18* (3–4), 253–269. doi:10.1016/S0923-4748(01)00037-6

Kakabadse, A., Kakabadse, N. K., & Kouzmin, A. (2003). Reinventing the democratic governance project through information technology? A growing agenda for debate. *Public Administration Review, 63* (1), 44–60. doi: 10.2307/977520

Korac-Kakabadse, N., Kouzmin, A., & Korac-Kakabadse, A. (2000). Information technology and development: Creating IT harems, fostering new colonialism or solving wicked policy problems? *Public Administration and Development, 20* (3), 171–184.

Kawabata, M., Thapa, R. B., Oguchi, T., & Tsou, M.-H. (2010). Multidisciplinary cooperation in GIS education: A case study of US colleges and universities. *Journal of Geography in Higher Education, 34* (4), 493–509. doi:10.1080/03098265.2010.486896

Kettl, D. F. (2002). *The Transformation of Governance: Public Administration for Twenty-First Century America.* Baltimore, MD: Johns Hopkins University Press.

Kraemer, K. L., & Dedrick, J. (1997). Computing and public organizations. *Journal of Public Administration Research and Theory, 7* (1), 89–112.

La Porte, T. R. (1971). The context for technology assessment: A changing perspective for public organization. *Public Administration Review, 31* (1), 63–73.

La Porte, T. R. (1994). A state of the field: Increasing relative ignorance. *Journal of Public Administration Research and Theory, 4* (1), 5–15.

Lamb, A., & Johnson, L. (2010). Virtual expeditions: Google Earth, GIS, and Geovisualization technologies in teaching and learning. *Teacher Librarian, 37* (3), 81–85.

Lardeau, M. P., Healey, G., & Ford, J. (2011). The use of photovoice to document and characterize the food security of users of community food programs in Iqaluit, Nunavut. *Rural Remote Health, 11* (2), 1680.

Latour, B. (1986). Visualization and cognition: Thinking with eyes and hands. *Knowledge and Society: Studies in the Sociology of Culture Past and Present, 6,* 1–40.

Latour, B. (1987). *Science in Action: How to Follow Scientists and Engineers through Society.* Cambridge, MA: Harvard University Press.

Latour, B. (2002). Morality and technology: The end of the means. *Theory, Culture & Society, 19* (5/6), 247–260.

Latour, B. (2005). *Reassembling the Social: An Introduction to Actor-Network-Theory.* New York, NY: Oxford University Press.

Lave, J., & Wenger, E. (1991). *Situated Learning: Legitimate Peripheral Participation.* New York, NY: Cambridge University Press.

Law, J. (1991). *A Sociology of Monsters: Essays on Power, Technology, and Domination.* London; New York: Routledge.

Law, J., & Callon, M. (1992). The life and death of an aircraft: A network analysis of technical change. In W. E. Bijker & J. Law (Eds.), *Shaping Technology/Building Society: Studies in Sociotechnical Change* (pp. 21–52). Cambridge, MA: MIT Press.

Lessig, L. (2011). *Republic, Lost: How Money Corrupts Congress—And a Plan to Stop It.* New York, NY: Twelve.

Lindstedt, C., & Naurin, D. (2010). Transparency is not enough: Making transparency effective in reducing corruption. *International Political Science Review, 31* (3), 301–322.

Lohr, S. (2012). The age of big data. *New York Times*. Retrieved from http://www.nytimes.com

MacKenzie, D. A., & Wajcman, J. (1999). *The Social Shaping of Technology* (2nd ed.). Philadelphia, PA: Open University Press.

MacKinnon, R. (2012). *Consent of the Networked: The Worldwide Struggle for Internet Freedom*. New York, NY: Basic Books.

Malamud, C. (2009, September 10). *By the People* [Video]. Retrieved from http://public.resource.org/people/

March, J. G. (1976). The technology of foolishness. In J. March & J. Olsen (Eds.), *Ambiguity and Choice in Organizations* (pp. 69–81). Bergen, Norway: Universitetsforlaget.

Marcuse, H. (1964). *One-Dimensional Man: Studies in the Ideology of Advanced Industrial Society*. Boston, MA: Beacon Press.

Markoff, J. (2007, August 20). A quest to get more court rulings online, and free. *New York Times*. Retrieved from http://nytimes.com

Marshall, G. (2012). Applying film to public administration. *Administrative Theory and Praxis, 34* (1), 133–142.

Martin, R. L., & Osberg, S. (2007). Social entrepreneurship: The case for definition. *Stanford Social Innovation Review, 5* (2), 28–39.

McAfee, A. (2003). When too much IT knowledge is a dangerous thing. *MIT Sloan Management Review, 44* (2), 83.

McGill, W. J. (2005). Moving local government GIS from the tactical to the practical. In C. Fleming (Ed.), *The GIS Guide for Local Government Officials* (pp. 9–28). Redlands, CA: ESRI Press.

McGuire, T., Manyika, J., & Chui, M. (2012, July/August). Why big data is the new competitive advantage. *Ivey Business Journal, 76* (4). Retrieved from http://www.iveybusinessjournal.com

McSwite, O. C. (2002). Narrative in literature, film, video, and painting: Theoretical and practical considerations of their relevance to public administration. *Public Voices, 5* (1–2), 89–96.

Mennis, J. (2003). Using Geographic Information Systems to create and analyze statistical surfaces of population and risk for environmental justice analysis. *Social Science Quarterly, 83* (1), 281–297. doi:10.1111/1540-6237.00083

Miller, H. T. (2012). *Governing Narratives: Symbolic Politics and Policy Change*. Tuscaloosa, AL: University of Alabama Press.

Miller, H. T., & Jaja, C. (2005). Some evidence of a pluralistic discipline: A narrative analysis of public administration symposia. *Public Administration Review, 65* (6), 728–738. doi:10.1111/j.1540-6210.2005.00501.x

Minsky, M. L. (1985). *Robotics*. Garden City, NY: Anchor Press/Doubleday.

Minsky, M. L. (1986). *The Society of Mind*. New York, NY: Simon and Schuster.

Minsky, M. L., & Papert. S. (1973). *Artificial Intelligence*. Eugene, OR: Oregon State System of Higher Education.

Monmonier, M. S. (1991). *How to Lie with Maps*. Chicago: University of Chicago Press.

Morgan, G. (1997). *Imaginization: New Mindsets for Seeing, Organizing, and Managing.* Thousand Oaks, CA: Berrett-Koehler.

Neill, C., Leipert, B. D., Garcia, A. C., & Kloseck, M. (2011). Using photovoice methodology to investigate facilitators and barriers to food acquisition and preparation by rural older women. *Journal of Nutrition in Gerontology and Geriatrics, 30* (3), 225–247. doi:10.1080/21551197.2011.591268

Nespor, J., & Barylske, J. (1991). Narrative discourse and teacher knowledge. *American Educational Research Journal, 28* (4), 805–823. doi:10.3102//00028 312028312028004805

Office of Management and Budget. (2011). *Capital Programming Guide: Planning, Budgeting, and Acquisition of Capital Assets.* (OMB Circular A-11). Retrieved from http://www.whitehouse.gov/sites/default/files/omb/assets/a11 _current_year/capital_programming_guide.pdf

Ogburn, W. F., & Thomas, D. (1922). Are inventions inevitable? A note on social evolution. *Political Science Quarterly, 37* (1), 83–98.

Orlikowski, W. J. (1991). *The Duality of Technology: Rethinking the Concept of Technology in Organizations.* Cambridge, MA: MIT.

Orlikowski, W. J. (2000). Using technology and constituting structures: A practice lens for studying technology in organizations. *Organization Science, 11* (4), 404–428. doi:10.1287/orsc.11.4.404.14600

Orlikowski, W. J., & Robey, D. (1991). Information technology and the structuring of organizations. *Information Systems Research, 2* (2), 143–169. doi:10.1287/ isre.2.2.143

Ospina, S. M., & Dodge, J. (2005a). It's about time: Catching method up to meaning—the usefulness of narrative inquiry in public administration research. *Public Administration Review, 65* (2), 143–157. doi:10.1111/j.1540-6210.2005.00440.x

Ospina, S. M., & Dodge, J. (2005b). Narrative inquiry and the search for connectedness: Practitioners and academics developing public administration scholarship. *Public Administration Review, 65* (4), 409–423.

Ospina, S., & Foldy, E. (2009). A critical review of race and ethnicity in the leadership literature: Surfacing context, power and the collective dimensions of leadership. *Leadership Quarterly, 20* (6), 876–896.

Ostrom, E. (2000). Collective action and the evolution of social norms. *Journal of Economic Perspectives, 14* (3), 137–158. doi:10.1257/jep/14/3/137

Oxford University, P. '2000). *Oxford English Dictionary.* [Oxford, England]: Oxford University Press.

Pentland, B. T., & Feldman, M. S. (2008). Designing routines: On the folly of designing artifacts, while hoping for patterns of action. *Information and Organization, 18* (4), 235–250. doi:10.1016/j.infoandorg.2008.08.001

Podolny, J. M. (2011). A conversation with James G. March on learning about leadership. *Academy of Management Learning & Education, 10* (3), 502–506. doi:10.5465/amle.2011.0003

Polletta, F., & Lee, J. (2006). Is telling stories good for democracy? Rhetoric in public deliberation after 9/11. *American Sociological Review, 71* (5), 699–721. doi:10.1177/000312240607100501

Rabassa, G. (2005). *If This Be Treason: Translation and Its Discontents.* New York, NY: New Directions.

Riccucci, N. (2010). *Public Administration: Traditions of Inquiry and Philosophies of Knowledge.* Washington, DC: Georgetown University Press.

Rohr, J. A. (1986). *To Run A Constitution: The Legitimacy of the Administrative State.* Lawrence: University of Kansas Press.

Rohr, J. A. (1989). *Ethics for Bureaucrats: An Essay on Law and Values.* New York: M. Dekker.

Rokeach, M. (1968). *Beliefs, Attitudes, and Values: A Theory of Organization and Change.* San Francisco: Jossey-Bass.

Rokeach, M. (1973). *The Nature of Human Values.* New York, NY: Free Press.

Rokeach, M. (1979). *Understanding Human Values: Individual and Societal.* New York, NY: Free Press.

Scheibel, D. (2003). Reality ends here: Graffiti as an artifact. In R. P. Clair (Ed.), *Expressions of Ethnography: Novel Approaches to Qualitative Methods* (pp. 219–230). Albany, NY: State University of New York Press.

Selznick, P. (1957). *Leadership in Administration: A Sociological Interpretation.* Evanston, IL: Row Peterson.

Severson, K., & Brown, R. (2011, May 9). Mayor's world remade in an instant. *New York Times.* Retrieved from http://www.nytimes.com/2011/05/10/us/10voices.html

Shannon, C. E., & Weaver, W. (1964). *The Mathematical Theory of Communication.* Urbana, IL: University of Illinois Press.

Sliwinski, A. (2002). Spatial point pattern analysis for targeting prospective new customers: Bringing GIS functionality into direct marketing. *Journal of Geographic Information and Decision Analysis, 61* (1), 31–48.

Sowell, T. (1980). *Knowledge and Decisions.* New York: Basic Books, Inc.

Sowell, T. (2002). *A Conflict of Visions: Ideological Origins of Political Struggles.* New York, NY: Basic Books.

Spicer, M. (1995). *The Founders, the Constitution, and Public Administration: A Conflict in Worldviews.* Washington, DC: Georgetown University Press.

Spicer, M. (2007). Politics and the limits of a science of governance: Some reflections on the thought of Bernard Crick. *Public Administration Review, 67* (4), 768–779. doi:10.1111/j.1540-6210.2007.00759.x

Spicer, M. (2010). *In Defense of Politics in Public Administration: A Value Pluralist Perspective.* Tuscaloosa, AL: University of Alabama Press.

Stivers, C. (2002). *Gender Images in Public Administration: Legitimacy and the Administrative State* (2nd ed.). Thousand Oaks, CA: Sage.

Stivers, C. (2008). *Governance in Dark Times: Practical Philosophy for Public Service.* Washington, DC: Georgetown University Press.

Stivers, C. (2009). The ontology of public space grounding governance in social reality. *American Behavioral Scientist, 52* (7), 1095–1108. doi:10.1177/0002764208327677

Stivers, C. (2011). Administration and the limits of democracy: The space of 19th-century American governance. *Administration & Society, 43* (6), 623–642. doi:10.1177/0095399711416079

Strum, S., & Latour, B. (1999). Redefining the social link: From baboons to humans. In D. A. MacKenzie & J. Wajcman (Eds.), *The Social Shaping of Technology* (pp. 116–125). Philadelphia, PA: Open University Press.

Swidler, A. (1986). Culture in action: Symbols and strategies. *American Sociological Review, 51* (2), 273–286.

Tanjasiri, S. P., Lew, R., Kuratani, D. G., Wong, M., & Fu, L. (2011). Using Photovoice to assess and promote environmental approaches to tobacco control in AAPI communities. *Health Promotion Practice, 12* (5), 654–665. doi:10.1177/1524839910369987

Taylor, J. L., & Rymer, J. T. (2011, July). *The Chief Financial Officers Act of 1990—20 Years Later.* Chief Financial Officers Council & Council of the Inspectors General on Integrity and Efficiency. Retrieved from http://www.whitehouse.gov/sites/default/files/omb/financial/cfo-act-report.pdf

Terry, L. D. (1995). *Leadership of Public Bureaucracies: The Administrator as Conservator.* Thousand Oaks, CA: Sage.

Thompson, J. D. (1967). *Organizations in Action: Social Science Bases of Administrative Theory.* New York, NY: McGraw-Hill.

Thompson, J. D. (2003). *Organizations in Action: Social Science Bases of Administrative Theory.* New Brunswick, NJ: Transaction.

Thrall, G. I. (2002). *Business Geography and New Real Estate Market Analysis.* New York, NY: Oxford University Press.

Timberg, C., & Gellman, B. (2013, August 29). NSA paying U.S. companies for access to communications networks. *Washington Post.* Retrieved from http://www.washingtonpost.com

Tippett, K. (2010, November). *Reconnecting with Compassion* [video lecture]. Retrieved from http://www.ted.com/talks/krista_tippett_reconnection_with_compassion

Turow, J. (2011). *The Daily You: How the New Advertising Industry Is Defining Your Identity and Your Worth.* New Haven, CT: Yale University Press.

UN World Food Programme. (2015). *Hunger Statistics: United Nations World Food Programme.* Retrieved from http://www.wfp.org/hunger/stats

US Government. (2011). Federal agency participation—Data.gov. Retrieved December 2011 from http://www.data.gov/metric

Wang, C. (1999). Photovoice: A participatory action research strategy applied to women's health. *Journal of Women's Health, 8* (2), 185–192. doi:10.1089/jwh.1999.8.185

Wang, C., & Burris, M. A. (1994). Empowerment through photo novella: Portraits of participation. *Health Education Quarterly, 21* (2), 171–186. doi:10.1177/109019819402100204

Washington, G. (2008). *Washington's Farewell Address 1796*. Retrieved from Yale University Lillian Goldman Law Library Avalon Project website: http://avalon.law.yale.edu/18th_century/washing.asp

Weick (1979). *The Social Psychology of Organizing*. Menlo Park, CA: Addison-Wesley.

Wellman, B., Quan-Haase, A., Boase, J., Chen, W., Hampton, K., Díaz, I., & Miyata, K. (2003). The social affordances of the Internet for networked individualism. *Journal of Computer-Mediated Communication, 8* (3). doi:10.1111/j.1083–6101.2003.tb00216.x

White, J. D. (2007). *Managing Information in the Public Sector*. Armonk, NY: M. E. Sharpe.

Wildavsky, A. B. (1987). *Speaking Truth to Power: The Art and Craft of Policy Analysis*. New Brunswick, NJ: Transaction.

Wilson, W. (1887). The study of administration. *Political Science Quarterly, 2* (2), 197–222.

Yunus, M., & Weber, K. (2007). *Creating a World without Poverty: Social Business and the Future of Capitalism*. New York, NY: Public Affairs.

Yunus, M., & Weber, K. (2010). *Building Social Business: The New Kind of Capitalism That Serves Humanity's Most Pressing Needs*. New York, NY: Public Affairs.

Index

Page numbers in *italics* indicate figures.

census data, 96
CERN (European Organization for Nuclear Research), 1
charismatic views of computer technology, 49–51
Chief Financial Officer Act, 46
citizen commodification, xi
citizens: action to impact lives of, 7–8; data collection from, 118–19; disconnect between government and, 34; informal knowledge and, 28; information asymmetry and, 104–5; interactivity among, 81–82; online interactions with government, 97; surveillance of, 93, 105, 107, 108; trust in government, 7, 122–24
civil society enterprises, 123–24
classification of information: as formal and informal knowledge, 28–30, *29*; overview of, 26–28; public policy making and formal information, 32–34; theory of information flow, 30–32
codification of values, 79
coding errors, 96–97
Coleman, Stephen, 68
commercialization of federal systems, 46
communication: constitutive feature of, 120; face-to-face, 124. *See also* information and communication technologies
community assessment tool, photovoice method as, 70
context: bringing to classroom, 81; building information from, using IT, 63; operational meaning of, 62
contextual information, 120, 121
contextualization of information. *See* information contextualization
Cooper, Terry, 73
Critical Theory of Technology (Feenburg), 41
culture and technology, 47–51
"Culture in Action" (Swidler), 48, 49

Dandeker, Christopher, 130n4
Dashboard (itdashboard.gov), 45
data: coding errors in, 96–97; inferences from, 100–101, 112; information compared to, 23–24; learning from, 90; machine intelligence to process, 98–103; in mathematical theory of communication, 24–26; meaning of, 23–24; noise of, 25, 112–13, 114; privacy of, 93, 97, 99; public, 92–94, 96; as ruling decision making, 7. *See also* data collection
data-based learning, xi, xii
data collection: from citizens, 118–19; competitive advantage over, xi; need for management and protection of, x–xi; on program success and failure, 108–13; scenarios of insatiable thirst for, 11–15
data.gov portal, xiii, 72, 105
data mining: of public information, 32; role of, 91–98
Davidson, C. N., 81, 84, 101
dead routines, 54–55
decision making: contextual information and, 120–21; GIS in, 71–72; as science and art, 73–78, 80–81
Dedrick, J., 43
democracy: balance between practice and administration in, 121–22; data mining in, 93; digital connectedness as right in, 83–84; diversity of values and, 63; ethics of technology in public institutions of, 88; information as fundamental to, 5–6; information asymmetry and, 104–5; instrumental view of technology and, 41–42; public administrator role in, 103
democratic practice, organized action of public agencies that constrain, 118–19
Denrell, J., 114
Dewey, John, 28
digital narrative, 67–68

Heaton, B., 99
Heidegger, Martin, 35, 38, 39, 52
Homeland Security Act of 2002, 91
"hot-stove effect," 114
how-to-do information, 2–3, 7
how-to-do question, 116
human actors and technology, xv
human experience and technology, 36–42
humanism, 74–75
human relations technology enactment, 56, 56–57
Hummel, R. P., 67
hunger and poverty, 84

ICT. See information and communication technologies
Ihde, Don, *Technology and the Lifeworld*, 37
immutability: of artifact, 53–54; of technology enactment, 56–57
individualism, 49
individual rationalism, 46
inferences from data, 100–101, 112
inform, definition of, 22
informal information: overview of, 3–7; practice and, 9, *10*, 11; for understanding values, 61–62. *See also* information contextualization
informal knowledge, 28–30, *29*
information: categories of, 2–3; classification of, 26–34, *29*; as constituted from practice, 107–15; disconnection between action and, 84; duality of, 3–7, 9–10, *10*; as measurable and contextual, 120; power and, 120–21; principles of, 22–26; translation of, 52, 69, 115–16. *See also* formal information; informal information; information contextualization; theory of information; traces of information; value of information
Information Age: practice-administra-

tion dichotomy of, ix–x, 115–25, *119*; public administration in, 11–15, *17*
informational resources, 55–56, 57
information and communication technologies (ICT): as charismatic, 49–51; formal information and, 3–4; impact of on human progress, 83–84; as mediators, 65; as preconceived for desired outcome, xiv; service provision and, ix; uses of, x–xi, xii
information contextualization: geographic information system, visualization, and, 70–72; Google Earth, Bing, social media mash-up, and, 72–73; implications of, 78–81; narrative inquiry and, 66–69; overview of, 61–62; photovoice method and, 69–70; science, art of decision making, and, 73–78; structuration theory and, 62–65, 74; tool kit for, 66–73
information overflow, 21, 91
information processing, technology in, xiii–xiv
information technology (IT): building information from context using, 63; duality of information and, 6; failures of projects in, 86–87; limitations of, 32–33; as materializing imagination, 64–65; overview of, 35–36; percent of successful projects in, xii; as tool, xii
inscription, 52–53, 54
institutional value commitment and leadership, 85–90
institutions: development of organizations into, 85; management systems for, 86; as techno-centric, 87
instrumental rationality: overview of, 42–44; private sector influence and, 44–46
instrumental view: of public admin-